MW00562045

GUIDE TO PREPARING THE CORPORATE QUALITY MANUAL

QUALITY AND RELIABILITY

A Series Edited by

EDWARD G. SCHILLING
Coordinating Editor
Center for Quality and Applied Statistics
Rochester Institute of Technology
Rochester, New York

RICHARD S. BINGHAM, JR.
Associate Editor for
Quality Management
Consultant
Brooksville, Florida

LARRY RABINOWITZ
Associate Editor for
Statistical Methods
College of William and Mary
Williamsburg, Virginia

THOMAS WITT
Associate Editor for
Statistical Quality Control
Rochester Institute of Technology
Rochester, New York

1. Designing for Minimal Maintenance Expense: The Practical Application of Reliability and Maintainability, *Marvin A. Moss*
2. Quality Control for Profit: Second Edition, Revised and Expanded, *Ronald H. Lester, Norbert L. Enrick, and Harry E. Mottley, Jr.*
3. QCPAC: Statistical Quality Control on the IBM PC, *Steven M. Zimmerman and Leo M. Conrad*
4. Quality by Experimental Design, *Thomas B. Barker*
5. Applications of Quality Control in the Service Industry, *A. C. Rosander*
6. Integrated Product Testing and Evaluating: A Systems Approach to Improve Reliability and Quality, Revised Edition, *Harold L. Gilmore and Herbert C. Schwartz*
7. Quality Management Handbook, *edited by Loren Walsh, Ralph Wurster, and Raymond J. Kimber*

ADDITIONAL VOLUMES IN PREPARATION

GUIDE TO PREPARING THE CORPORATE QUALITY MANUAL

Bernard Froman
*Conseil en Management
et Assurance de la Qualité
Paris, France*

Translation by

Tony Pierce

MARCEL DEKKER, INC.   NEW YORK · BASEL · HONG KONG

Library of Congress Cataloging-in-Publication Data

Froman, Bernard
   [Manual Qualité. English]
   Guide to preparing the corporate quality manual / Bernard Froman;
translation by Tony Pierce.
      p.   cm. — (Quality and reliability; 51)
   Translation of: Le Manual Qualité.
   Includes index.
   ISBN 0-8247-0090-2 (hardcover: alk. paper)
      1. Quality control. 2. Total quality management. 3. ISO 9000
Series Standards. 4. Reliability (Engineering) I. Title.
II. Series.
   TS156.F76513   1997
   658.5'62—dc21

                                                            97-6373
                                                               CIP

The contents of this volume were originally published as *Le Manuel Qualité: Outil Stratégique d'une Démarche Qualité*, 2nd Edition, by Bernard Froman. Copyright © 1995 by AFNOR, Tour Europe, 92049 Paris la Défense cedex, France.

The publisher offers discounts on this book when ordered in bulk quantities. For more information, write to Special Sales/Professional Marketing at the address below.

This book is printed on acid-free paper.

MARCEL DEKKER, INC.
270 Madison Avenue, New York, New York 10016
http://www.dekker.com

Current printing (last digit):
10 9 8 7 6 5 4 3 2 1

PRINTED IN THE UNITED STATES OF AMERICA

About the Series

The genesis of modern methods of quality and reliability will be found in a sample memo dated May 16, 1924, in which Walter A. Shewhart proposed the control chart for the analysis of inspection data. This led to a broadening of the concept of inspection from emphasis on detection and correction of defective material to control of quality through analysis and prevention of quality problems. Subsequent concern for product performance in the hands of the user stimulated development of the systems and techniques of reliability. Emphasis on the consumer as the ultimate judge of quality serves as the catalyst to bring about the integration of the methodology of quality with that of reliability. Thus, the innovations that came out of the control chart spawned a philosophy of control of quality and reliability that has come to include not only the methodology of the statistical sciences and engineering, but also the use of appropriate management methods together with various motivational procedures in a concerted effort dedicated to quality improvement.

This series is intended to provide a vehicle to foster interaction of the elements of the modern approach to quality, including statistical applications, quality and reliability engineering, management, and motivational aspects. It is a forum in which the subject matter of these various areas can be brought together to allow for effective integration of appropriate techniques. This will promote the true benefit of each, which can be achieved only through their interaction. In this sense, the whole of quality and reliability is greater than the sum of its parts, as each element augments the others.

The contributors to this series have been encouraged to discuss fundamental concepts as well as methodology, technology, and procedures at
the leading edge of the discipline. Thus, new concepts are placed in
proper perspective in these evolving disciplines. The series is intended
for those in manufacturing, engineering, and marketing and management, as well as the consuming public, all of whom have an interest and
stake in the products and services that are the lifeblood of the economic
system.

The modern approach to quality and reliability concerns excellence:
excellence when the product is designed, excellence when the product is
made, excellence as the product is used, and excellence throughout its
lifetime. But excellence does not result without effort, and products and
services of superior quality and reliability require an appropriate combination of statistical, engineering, management, and motivational effort.
This effort can be directed for maximum benefit only in light of timely
knowledge of approaches and methods that have been developed and
are available in these areas of expertise. Within the volumes of this
series, the reader will find the means to create, control, correct, and
improve quality and reliability in ways that are cost effective, that
enhance productivity, and that create a motivational atmosphere that is
harmonious and constructive. It is dedicated to that end and to the
readers whose study of quality and reliability will lead to greater
understanding of their products, their processes, their workplaces, and
themselves.

Edward G. Schilling

Foreword

From many years of industrial, academic and consulting experience, I have developed a deep appreciation of the need for well-structured, informative and pragmatic documentation of a company's quality-related activities and their impact on the improvement of the enterprise.

Historically, a new business must blaze a path for its growth. The start is often bumpy and one from which learning comes from doing. Sometimes a company policy is formulated early on; other times it evolves from practices and procedures that have been accepted as a good way of functioning. As an enterprise grows, its operations becomes more complex, often with little time for the documentation of its "way of life." As a result, there are companies that must struggle to comply with requests from customers, suppliers and auditors for written evidence (a manual) that the company is sufficiently organized to satisfy expectations (particularly those dealing with quality, safety and the environment) that arise internally and/or externally.

The *Guide to Preparing the Corporate Quality Manual* is just that: an indispensable guide! It has been prepared with due care and exhibits a precision that derives from the practical experience of its author. First of all, it is an effective prompt for self-assessment - "Know thyself" - and it is a useful instrument for informing prospective customers and auditors about the company's policies, practices and procedures that are designed to result in customer satisfaction. Moreover, it provides customer suppliers with a meaningful basis for demonstrating an approach to attainable quality assurance.

The quest for compliance with ISO 9000 Quality System Standards, national quality awards criteria, and above all, customer requirements has brought pressure to bear on companies to create well-constructed manuals that describe what they do internally to address supplier–processor–customer relationships that lead to superior performance.

For those who have yet to prepare a corporate quality manual or for those who desire to improve existing ones, I heartily recommend the use of the guidelines offered in Mr. Froman's thorough, scholarly work.

> *John D. Hromi, Dr. Eng.*
> *Professor Emeritus*
> *The John D. Hromi Center for Quality*
>   *and Applied Statistics*
> *Rochester Institute of Technology*
> *President (1981–1982),*
> *American Society for Quality Control*

Preface

The quality manual is the "written image of company." It describes how the organization should set up and comply with its own quality policy:
- in house, it is the reference document used for internal quality management;
- in customer–supplier relations, it is testimony to the quality assurance level that has been achieved and that is used to win the customer's confidence.

The development of quality system certification makes the quality manual an indispensable tool because it is the document on which a firm's aptitude evaluation audit is based. However, the manual also concerns firms that use different criteria, such as TQM and the theories of Malcolm Baldrige. In practice, the quality head of every organization has to develop a quality manual at one time or another. More generally, the quality manual is a powerful tool for achieving progress; its application is gradually spreading to various new areas, such as environmental protection and safety.

So why is the quality manual so important? It is not just because the 1994 revision of the ISO 1900 standards requires it. If it is written clearly and simply, not only does the quality manual give an accurate picture of the company, but it also serves as a definite strategic tool in terms of quality management, offering companies all the obvious commercial and financial benefits that accrue from improved organization.

The essential objectives of this book are as follows:
- to cast some light on quality management concepts in compliance with those of the ISO 9000 family of standards,
- to outline a basic scheme for thought and decision regarding the choice of quality policy in the application of these concepts,
- to specify the essential role of the quality manual among a firm's documents for the implementation and development of a quality system,
- to offer clear and pragmatic advice on how to make the quality system efficient and how to describe that system in the quality manual.

So how do we go about choosing the structure and content of a quality manual so that it ties in with the adopted quality policy, while being appropriate for the structure and size of the firm, its customers, and its certification goals and still managing to assure the internal improvement of the organization and its processes? Even for quality professionals, the task is not always easy; it is certainly far less easy when it is being done for the first time.

In the first part of the book, I define the essential role of the quality manual as part of the firm's documents. Then, referring to the main concepts developed in the ISO standards, I give helpful advice with practical examples regarding the structure chosen. This choice is obviously important because some firms simply make do with compiling a manual by adding to the ring-binder a first section called "Quality Policies," followed by another titled "Procedures and Process Instructions or Forms," while keying the subsequent sections to the quality elements of the prevailing ISO standard. Others prefer to analyze the different processes and describe them, then choose an appropriate structure. This may be a better way to build a quality manual as **a strategic tool for an approach to quality** because it encompasses quality models other than those of ISO 9000—TQM, for instance.

In the second part, I refer back to the various elements of a quality system and explain how they can be used in an organization by using simple examples drawn from real-life situations encountered in various companies.

This book is an endeavor to address fully the needs of all those seeking an approach to documenting quality. It aims at taking some of the mystery out of quality management concepts and attempts to help explain the overall ISO 9000 standards philosophy and other quality models, without going into tedious detail involving the interpretation of several different standards. Its unique characteristic is that of combining different quality policy guidelines, based on the most advanced quality concepts recognized worldwide, offering assistance, by means of practical examples, to those wishing to write a quality manual.

Its primary audience is managers who are confronted by the need to choose a quality policy, engineers or quality professionals who have to implement that policy and quality consultants who need to guide companies in the writing or improvement of their quality manuals or helping them in their efforts towards certification.

This book was conceived on the basis of more than 15 years' experience as quality assurance manager at major nuclear corporations and 6 years as Chairman of ISO/TC 176/SCI—"Quality Management and Quality Assurance—Concepts and Terminology" (1987–1993). I also took advantage of experience gained from observing the work carried out on the quality manual by Mr. Lawrence A. Wilson, U.S. Convener of ISO/TC 176/SC3WG2 (development of ISO 10013). My work as expert with two certification organizations (AFAQ and ASCERT International) has also been a valuable source of material. I have endeavored to draw upon manuals from both small firms and major corporations.

I would like to thank Mr. Patrick Bertin, Quality Director of Rhône-Poulenc and Chairman of AFAQ Chemicals and Oil Certification Committee; Mr. Michel Vallès, Head of CERIB's Division of Quality and Industrial Productivity, and Chairman of AFAQ's Multi-Area Committee; and Mr. Paul Goossens, Chargé de Mission with the

Commission Centrale des Marchés, who agreed to share the fruit of
their great on-site experience and to assist with the re-reading of
certain chapters. I would also particularly like to thank Mr. Hervé
David, Secretary of AFNOR's General Committee on Quality and
Management and of the ISO Sub-Committee on Concepts and
Terminology, who, by his close collaboration within the ISO/TC176
Committee, has always provided the best advice.

When this book was first published in French, the ISO 8402 standard
of 1986 and the ISO 9000 series of standards of 1987 (9000, 9001,
9002, 9003, 9004) were being revised. The present edition takes into
account the amendments contained in the 1994 revision of the
following standards: ISO 8402, 9000-1, 9001, 9002, 9003 and 9004-1.
Furthermore, it should be noted that these international standards have
been updated as European and French standards NF EN ISO 8402 and
NF EN ISO 9000. The reader should therefore check the date of
validity of the standards being referred to.

Bernard Froman

Contents

List of Figures

Introduction

Quality

The quality objective is to meet stated and implied needs. These needs can be those of:

- the product or service users (already defined or to be identified),
- the requirements of society (obligations governed by laws, regulations, codes or other considerations, aiming at, for example, safety, environmental protection, health, etc.),
- the requirements for good internal organizational management (internal needs defined by management).

This objective is addressed by implementing measures designed for the following purposes:

- for a product or service, to translate these needs into specifications as needed to implement them and to make sure that manufacturing conforms to the established specifications,
- for the organization's regulations, to ensure conformity with these regulations,
- for good internal management, to set up appropriate management tools.

Over the last few decades, the subjects of "product quality" and "quality inspection" have become extremely important, especially among technicians; moreover, the two phrases have become almost inseparable.

Technical evolution, combined with automation and the computerization of manufacturing methods have caused the moving of some control stages, previously spread throughout the range of operations, to the start or end of the manufacturing process. Naturally, in-process inspection - even if less thorough - is still used as a means of avoiding drifting from acceptable quality standards and is more and more often the subject of a contractual arrangement between customer and supplier.

In the concept of quality assurance appeared around 1950 and was first applied to nuclear and space programs, then, in more recent years, was extended to more diversified areas, especially following the success of the ISO 9000 family of standards. A new form of customer/supplier relationship was reached, based on the confidence of the customer towards the supplier and for which quality system certification was developed.

Quality management

The evolution of these quality concepts and their increasing application within different industrial and administrative areas have led to quality being recognized as a good management tool:
- real time quality data processing and analysis: inspection results, identification of production defects or organizational weaknesses, quality of supplies, etc., have enabled corrective actions to be taken each time an anomaly is reported;
- a change in the personnel's state of mind as a result of management's new awareness, intensified training programs, the introduction of management representatives in charge of quality, etc., have led to personnel being made more aware of their responsibilities and to the improvement of economic performance.

Gradually there has been a move from the simple concept of product quality inspection, to product or service quality assurance, and finally to quality management, which is the management activity that is devoted to quality in accordance with the ISO 9000 family of standards. This has involved creating and maintaining a state of mind that is just as appropriate in services and administrative environments as it is in more traditional production and manufacturing concerns.

Function and importance of the quality manual

It can be said that the quality manual is an "annotated picture" of the entity concerned (firm, administration etc.) in terms of its quality policy and its implementation of that policy. All other subsequent quality documents refer back to this manual. The work of preparing this manual is therefore the fundamental stage of any quality management approach.

For internal use, the manual is a basic quality management document that should be used as a reference at all hierarchical levels to achieve the objective of good management.

For use in the customer/supplier relationship (where it is often referred to as the "quality assurance manual"), it conveys the initial image of the supplier as far as quality assurance is concerned, and it is the first document used by to gain the confidence of the customer.

For use in quality system certification, it is the reference document on which the aptitude evaluation audit will be based.

The way the quality manual is drafted is most important as far as quality is concerned because it gives a picture of the level reached within the quality management approach; this level can be confirmed and even improved upon by its on-site application.

Finally, as a "living" document that can be modified in response to current developments, the importance of the quality manual can be fully realized if its field of application, and therefore its content and title, is extended to cover not only quality but also safety, and environmental management.

PART ONE - GENERAL PRINCIPLES

1 The Concepts of Quality Management and Quality Assurance

During the last few years, quality concepts have evolved a great deal (see introduction); therefore, it is important, **before working on the preparation of the quality manual**, that within the context of an organization's quality approach **everyone agree on well-defined concepts and terminology** so that users of the manual may understand clearly its content. It is best to refer to the most recent definitions as these have been the object of international consensus.

The principal concepts and the terminology currently retained by the international and French standards are contained in the ISO 8402 standard, the 1994 revision of which is translated into the European and French NF EN ISO 8402. The NF X 50-125 French standard, which contains additions to ISO 8402, will replace the French NF X 50-120 standard.

1.1 Quality

Quality is the "totality of characteristics of an **entity** that bear on its ability to meet stated and implied needs" (see ISO 8402).

In this revised standard, the word **"entity"** refers not only to a product or service as in the previous definition, but also to an activity, a process, an organization or a person. The word **"product,"** considered as the result of the activities and processes, can also refer to raw materials, hardware, software, services, etc.

The **satisfaction of the various needs** (see Introduction) implies that quality is the goal set throughout the industrial process (in the case of a product), or tertiary process (commerce or administration) or **throughout the life cycle of the product**: design, production, maintenance, etc. (see lower part of figure 1.1).

The introduction of the revised ISO 8402 standard broadly explains the quality concept and its different aspects.

1.2 Quality control and quality inspection

Quality control is defined as follows: "Operational techniques and activities that are used to fulfill requirements for quality" (see ISO 8402). These can be operational actions that can be applied to a pilot process (manufacturing progress, successive stages of a service) or to the elimination of nonconformities or deviations throughout a process in order to achieve the results expected.

Inspection is a quality control operation undertaken at a given stage of the process in question, to determine if, at that stage, the results obtained are in accordance with the specified requirements.

The quality control operations are the concern of the operational hierarchy, which has the responsibility for obtaining quality throughout this process (see figure 1.1).

1.3 Quality assurance

Quality assurance (often abbreviated as "QA") refers to "all the planned and systematic activities implemented within the 'quality system' (see ISO 8402) and, demonstrated as needed, to provide adequate confidence that an entity will fulfill requirements for quality" (see ISO 8402). Quality assurance consists of:

- **A coordinated group effort** aimed at:
  - . building confidence inside the organization that quality is being obtained.
  - . giving the customer (or regulatory authorities) confidence that quality is being obtained.
- **Planned and systematic actions** (provided for within a quality system environment) aimed at an organization's systematically stating in a document (such as a quality plan) the "control" operations that will be necessary to obtain quality.
- **Effective demonstration** of implemented actions through planned methods (documentation, quality audits, etc.). This demonstration must be based upon " objective evidence," and the " degree of demonstration " (see ISO 8402) is the extent to which evidence is produced, as needed, depending on technical complexity, safety etc. and cost criteria.
- **External auditing** of the organization's QA program. As a precondition, it is necessary to establish confidence within the organization via the organization's internal QA (implying internal audits); this must be one of the main objectives of the quality assurance manual (or quality manual, if it is assuming a dual role).

Although some quality control and quality assurance actions are interrelated, there must be no confusion between the two concepts:
- **quality control** relates to the fulfillment of quality requirements (operational or technical aspect) and,
- **quality assurance** aims at providing confidence that the requirements are fulfilled, within the organization (internal QA) and outside it to satisfy customers or regulatory and legislative authorities (external QA).

Therefore, in a quality system, a procedure can be a quality control document, because of the technical requirements that it contains, *as well as* a quality assurance document that will provide confi-

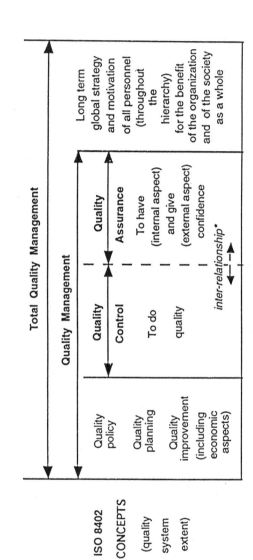

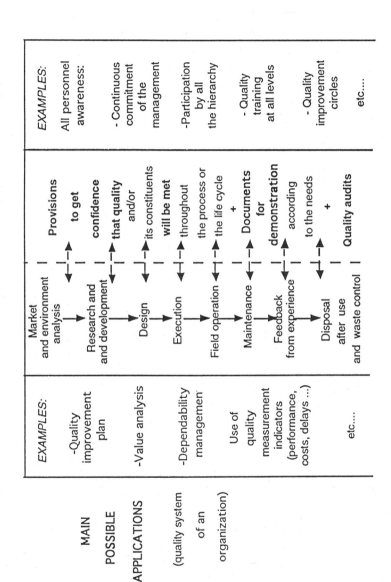

Figure 1.1 Constituents of quality management concepts

* *"inter-relationship"* :   Reminder of the inter-relations between quality control and quality assurance
to win confidence throughout the process or the life cycle

dence because of the complementary requirements it puts forth, which are aimed at demonstrating how quality will be achieved.

Quality assurance actions are generally carried out throughout a process, in a functional line, distinct from the quality control operational line, to meet the required independence that is necessary for securing confidence (see figure 1.1).

1.4 Quality management

Quality management is "all the activities of the overall management function that determine the quality policy, objectives and responsibilities and implement them by such means as quality planning, quality control, quality assurance and quality improvement, within the quality system" (see ISO 8402).

The way that management regards these aspects of quality reflects an organization's dedication to quality. Quality management consists, on one hand, of quality control and quality assurance and, on the other, of additional quality policy concepts such as quality improvement (including cost aspects). Quality management is applied to all the stages of a process or life cycle of a product or service; it can be extended to all the sections of an organization. It is implemented within or by an organization by setting up a quality system.

The **quality system** is, by definition, the "organizational structure, procedures, processes and resources needed to implement quality management" (see ISO 8402). It comprises therefore all the quality-related provisions that can be applied in full or in part by an organization and also throughout the life cycle of a product. In this sense, it can also be said that quality management "operates throughout the quality system" (see ISO 8402, Introduction).

Note that the term "quality assurance system," which is generally not an accepted meaning or definition, is to be avoided.

Total Quality Management (or TQM) is an extension of the quality management concept, as it pertains to the participation and motivation of all the members of the organization (from the top to the bottom of the hierarchy) in its own interest and for the benefit of its environment. Depending on the areas and on the different cultures, this concept can have various forms and designations. For example, **total quality**, "company-wide quality control" (CWQC), "management by total quality," or even designations of certain quality awards (Malcolm Baldrige, European Award (EFQM), French Quality Award). This term and its definition have been the object of an international consensus reported in the ISO 8402 standard (see § 3.7 and especially the corresponding note 5).

The relationship between these concepts and their possible application for an organization's quality system are shown in figure 1.1. Note that:
- the **vertical arrows** in the quality control column, which illustrate a series of elements in the "quality system" that can be the object of quality control and assurance provisions throughout the life cycle of a product (some are the object of standards - see ISO 9001 paragraph 4.4 for design, for example; others are not, or not yet, for example, field operation; certain organizations may find it useful or necessary to deal with these elements in their quality manuals);
- the **horizontal** "inter-relationship" **arrows,** which refer to the links between quality control and quality assurance throughout the process or life-cycle (to obtain confidence throughout the quality system).

For clarity's sake, these concepts are summarized in figure 1.2

Quality Management
(organization or enterprise)

Quality policy and quality objectives
Quality planning and quality improvement

Quality system
(organization - procedures - processes - resources)

Customer's needs for a product

Quality control (How to get it)	**Quality assurance** (Confidence in how to get it)
1. Anticipate what is to to be done	6. Demonstrate that the points 1, 2, 3, 4 and 5 have been complied with
2. Write down what is being planned	
3. Apply what has been written	7. Verify by audit that the quality system is adequate and that everything is processing as planed
4. Check andcorrect deviations	
5. Keep a record	8. Check the effect of corrective actions
Product performance	**Confidence in conformce**
Customer satisfaction	

Figure 1.2 Simplified pedagogical approach to the quality management concepts

2 The Quality Manual Within the Series of Quality Systems Documents

2.1 The quality documents pyramid

The documents relating to the "quality system" of an organization, i.e. relating to "organizational structure, procedures, processes and resources needed to implement quality management" (see ISO 8402), are:

- the quality (or quality assurance) manuals,
- the quality (or quality assurance) plans,
- the procedures,
- the various operational documents for the implementation of the quality system: instructions, operating methods management notes, specifications, forms, manufacturing and inspection process sheets, quality audit programs, minutes, certificates, reports, etc., and, more generally, all the records demonstrating how quality is obtained.

The nature of these documents, their importance in numbers, their typology and their user guide, depend of course on the nature of the organization (industrial group, small/medium sized enterprises/industries, production plant, service company, administration, etc.).

Quality (or QA) manuals are often represented at the top of the pyramid of the various quality documents because, by describing the organization's entire quality system in a very general manner, they refer to the overall existing quality implementation documents that they cover, i.e. the general procedures, or documents specific to various operations.

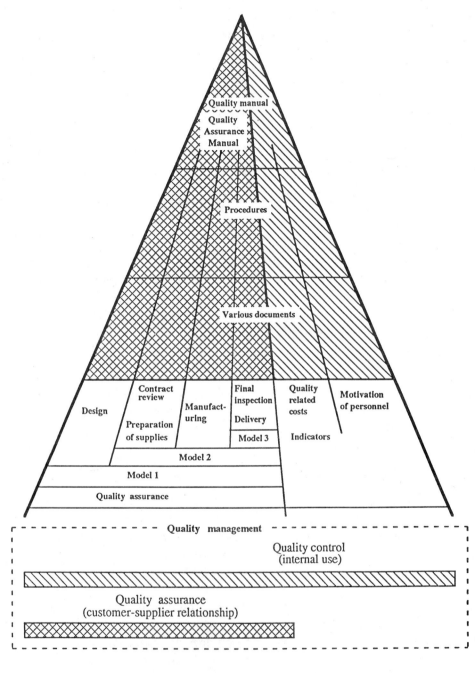

Figure 2.1   Quality documents pyramid

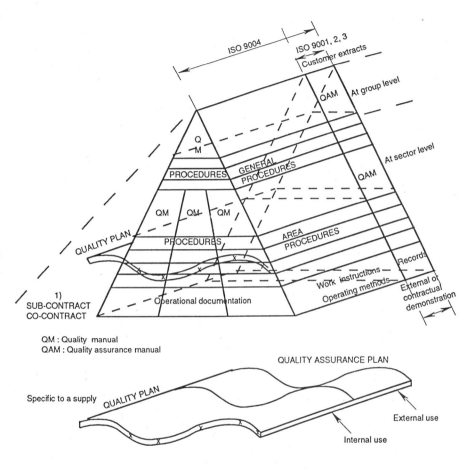

QM : Quality manual
QAM : Quality assurance manual

X Procedures for the product of the different areas concerned

1) Section of the quality plan concerned with sub-contract and co-contract

Figure 2.2 Example of breakdown and application of quality documents

The pyramid in figure 2.1 emphasizes the distinction between the quality control documents (internal, possibly confidential) and those on quality assurance (contractual).

The pyramid in figure 2.2 - extracted from the X 50 163 documentation sheet - applies principally to organizations made up of several branches and responsible for various products (requiring several quality manuals or quality assurance manuals) and reflects how a quality or quality assurance plan can be specific to a product within the documentary structure.

2.2 Quality manuals and quality plans

2.2.1 Quality manuals

The **quality manual** is a "document stating the quality policy and describing the quality system of an organization" (see ISO 8402 / X 50-163).

The **quality assurance manual** is a "document describing the general provisions applied by an organization with regard to quality assurance" (see X 50-162 and 163).

Why this differentiation?

Traditionally, there has been a distinction (where applicable) between documents for internal or confidential use and those for external use. The most recent French standards make a distinction between the "quality assurance manual," covering only the quality assurance, and the "quality manual," covering quality management and sometimes quality assurance.

But **this differentiation is optional**. Do not think that it is mandatory to draft two different manuals for internal and external use! Although it might well be preferable to differentiate them, one may be derived from the other by leaving out confidential elements or by adding internal management provisions.

For further details:

- **The term "quality manual"** is used, especially in the ISO 9000 family of standards for the organization's quality management (ISO 9004-1 or various quality approaches), for customer-supplier relationships and also for quality system certification. In the latter case, it has the same content as a quality assurance manual.

According to the X 50-162 documentation sheet and the X 50-163 documentation sheet, the quality manual refers to a document for the organization's internal use. It is created at the discretion of quality management and is not used as a basis for external audits or consulted by the customer. This distinction, also applied to the quality plan, is shown in figure 2.3 (extracted from the X 50-162 or 163 documentation sheets), taking into account the revised version of ISO 8402.

- **The term "quality assurance manual"** designates a document that may be contractually required, which may facilitate customer-supplier relations, more particularly by applying the "models for quality assurance" (see ISO 9001, 9002, 9003). It can be issued to a customer or organization dealing with quality system certification and is used as a basis for the corresponding external audits. In this case, it must not contain any confidential information.

Note that this evolution of French standards is in accordance with the revision of the ISO 8402 standard. Depending on the updated definitions of the X 50-163 documentation sheet - December 1992, a quality manual used for internal quality management can also be called a "quality management manual," as is also shown in figure 2.3. In this way, the term "quality manual" is used when the manual is to be used by the organization's quality management as well as for the customer-supplier relationships or for quality system certification.

Note also:

- that the quality assurance manual may contain extracts from the
  quality manual (CF X 50-163), if this manual has been drawn up and
  is maintained separately,
- that the quality manual, or if such is the case, the quality assurance
  manual, may be the only existing document applicable to the
  overall organization (see top part "covering" the pyramid in figure
  2.1).
- that the quality manual may also "cover" several quality manuals (or
  quality assurance manuals) established in the diverse branches or areas within the organization (see top part of the pyramid in figure
  2.2).

The choices in preparing and entitling one or several quality manuals
depends on the nature, size, structure and quality policy of the organization concerned (see § 3.1).

What other types of quality manuals are there?

Depending on industrial and regulatory concerns, the same manual
might group together, for example, the management of quality, safety
and environment, while drawing on the same documentation system
(see ISO 14000, paragraph 4.3.3).

2.2.2 Quality Plans

The **quality plan** is a "document setting out the specific quality practices, resources and sequence of activities relevant to a particular product, project or contract" (see ISO 8402).

The **quality assurance plan** is a "document describing specific quality assurance provisions taken by an organization to meet the specific
product and/or service requirements" (see X 50-163 and ISO X 50-
164).

Why this differentiation?

The distinction here is similar to the one made in the quality manuals and for the same reasons: to differentiate internal and confidential information from external information.

But, as is true for the manual, **this differentiation is optional**. While a quality plan is primarily destined to contribute to good quality management - as an internal need - a quality assurance plan may be required contractually to be submitted to the customer's audit. In this case also, the choice of title is optional, and if a need to distinguish between them occurs, drafting two very different internal and external plans is to be avoided. It is desirable that the quality assurance plan, if one has been drawn up, include extracts from the quality plan. This analogy is shown in figure 2.3.

In accordance with ISO standards, such a document should be referred to as the "quality management plan" when it results from a quality management internal approach.

Whereas the **quality manual** refers to the overall quality documents applicable within the organization or area that it covers, **the quality plan** refers only to the quality documents. For example, in the different areas involved with supply, the quality plan would cover only those procedures affecting the quality required in those areas. Finally, the quality plan can integrate previous quality plans established for each sub-area involved with the product, especially quality plans dealing with sub-contracting and co-contracting. These considerations are reflected in the pyramid of figure 2.2. Note that the plan "intersects" the different areas concerned that contribute to its making.

There exist other kinds of quality plans that can be called "strategic quality plans." These include different kinds of document established for planning the strategic quality objectives (see figure 2.3 - note 3 and chapter 8).

	Quality within the organization		
	Quality management within the organization ISO 9004		Customers-supplier relationships ISO 9001 - 9002 - 9003
Description of the organization's general provisions	Quality manual (1) (quality management manual) (2)		Quality assurance manual (2)
Description of the provisions specific to a supply	Quality plan (1) (Quality management plan) (2) (3)		Quality assurance plan (2)
	Documents for the organization's internal use established from their own free will		Documents that can be contractually required.

(1) Generic terminology used only in the ISO standards established before the revision of ISO 9000.
(2) Terminology introduced with the revision of ISO 8402 (Paragraph 3.12, note 3) and used for the revisions and drafts of ISO 9000 standards.
(3) A quality plan resulting from a "quality planning" activity (see ISO 8402 paragraph 3.3) may in some cases be the organization's "quality management plan", or may simply be a quality planning report for the use of the management. In this wider sense, a qualifier may be used, for example, "strategic quality plan". The term "quality plan" should be reserved for the specific meaning related to a product, project or particular contract (source: ISO/TC176, Toronto, May 1992).

Figure 2.3 Types of quality manuals and quality plans
(Source: ISO X 50-162 or 163 and ISO 8402)

Finally, as mentioned for the quality manuals in section 2.2.1, the provisions relating to quality, safety and care of the environment might well be grouped together within the same plan; this can lead to the concept, for example, of a "quality-safety plan" either as a general strategy or for a product or contract.

2.3 Procedures

A "procedure" is a "specified way to perform an activity" (see ISO 8402, paragraph 1.3).

As defined by these standards, "a written or documented procedure usually contains the purpose and scope of an activity; what shall be done and by whom; when, where and how it shall be done; what materials, equipment and document shall be used; and how this must be controlled and recorded."

It is therefore a group of rules written specifically for an activity, i.e., an entire organization or one of its areas: a workshop, a production or inspection operation, a quality assurance action, etc.

Types of procedures

The procedures are generally grouped together (sometimes taking the shape of a manual or folder), as follows:
- general procedures, applicable to the overall organization,
- procedures that relate specifically to different areas of the organization (see figure 2.2).
- where applicable, procedures that relate specifically to different current products or contracts.

The following can also be distinguished between:
- organizational procedures which form the general basis of the quality system: for example, all the requirements of a standard chosen from the ISO 9001, 2 or 3 series must be included;

- operational procedures detailing the technical and administrative activities; these can sometimes designate different operational documents such as working instructions, inspection methods, etc. (see § 2.4, which follows).

But remember, in some documents, the word "procedure" when used on its own, that is to say, without being qualified as "written" or "documented," can mean simply the "way to proceed." For requirements that have already been standardized and documented elsewhere, the manual should describe or refer to the various provisions that need to be taken to control an activity, with clear reference to the applicable standards.

The elements differentiating the quality manual from the procedures

Note that:
- a straightforward distinction is not always made between the quality manual, which must cover all the organization's quality system elements (those of the ISO 9000 standards family, for example), and the general procedures (i.e. "written" or "documented"): these are sometimes included in the quality manual or are added as attachments detailing, in some cases, the corresponding provisions of the application;
- other various quality documents - working instructions, organizational notes, organization charts, etc., - are sometimes uniformly presented under the word "procedure."

2.4 The different operational documents

No details will be given here of the types of quality documents that can form an organization's quality system; a general picture of this is provided at the beginning of this chapter and some are mentioned in figure 2.2. As a general rule, it is often possible to distinguish between:
- necessary preliminary documents:
    to define requirements: functional specifications and standards,

specifications, definition files,
.   to carry out activities: qualification documents, working instructions, management notes, operating methods, sheets, inspection plans.
-   reports and records necessary to demonstrate the achievement of the required quality: analysis or inspection forms or minutes, files established at the end of the manufacturing process, reports on quality meetings and visits, audit reports, management reviews, etc.;
-   usually at the request of some customers, open-ended documents such as some quality plans that are at first preoperative and then completed ("reviewed") during the successive inspections by an independant inspector or customer and that are to be included in the final report (see § 11.3.2 and 11.3.3);
-   organizational files: organization charts and assignments;
-   necessary training, development and maintenance files, etc.

The quality manual and the procedures with which it is concerned should refer to these various documents.

3 Principles for Writing a Quality Manual

To continue from the considerations made in chapter 2, first remember that the manual is a document describing, in a unified manner, the quality policy, the quality system and the organizational structure of the organization.

The following principles concern the production of a quality manual by an organization catering primarily to its internal quality management needs. But they can also be applied to a quality assurance manual that may be required contractually, and or they may even be suitable for the certification of the quality system (see § 2.2.1 and X 50-163 documentation sheet).

Depending on the desired objective, refer to the following standards:
- NF X 50-160: "Guide for the establishment of a quality manual" (approved standard),
- X 50-161: "Guide for drafting a quality assurance manual" (documentation sheet),
- X 50-162: "Guide for the establishing of the quality manual" (documentation sheet),
- ISO 10013: "Guidelines for developing quality manuals."

The principal reason for these standards is to facilitate establishing and administrating the quality manual and/or quality assurance manual so that they may be used effectively as an ongoing reference as part of an organization's quality system.

The following principles are designed to help the user of these standards make decisions with regard to the content and format of the manual. This should be done by taking into account their application and the need to adapt them appropriately, according to the specific needs of each organization.

3.1 Decision and function of top management

The decision to establish a quality or quality assurance manual to suit the quality approach adopted, by reflecting a progressive approach taking account of the economic, commercial and, in certain cases, regulatory environment of the organization, should be made at the highest level of management.

The decision must focus on the type of manual that needs to be established:

- If **no external constraint** exists (contractual or regulatory), the decision can be, for example, to establish a quality manual based on the applicable provisions of the ISO 9004 standard taking into account the activity concerned and using the X 50-161 documentation sheet to achieve good internal organizational management.
- If the decision abides by a **Total quality management or "total quality" approach** (see § 1.4), the quality manual may be similar to the previous one but will include additional provisions specific to the organization's own culture.
- If there is a need to answer **contractual or regulatory requirements** or to prepare the organization for the certification of its quality system, the manual will have to meet the requirements of one of the ISO 9001, 9002 or 9003 standards. This objective can be reached in different ways:
  - if no quality manual exists, a **quality assurance manual** may be drafted with the aid of the X 50-162 documentation sheet;
  - if there is a **quality manual** and if its content generally meets one of the above standards and is not confidential, it can be used, provided that its format is adapted to ease its use by auditors;

. if a quality manual exists and if it contains **confidential elements**, a **reduced version** of identical structure should be established. A **quality assurance manual** made of extracts from the quality manual itself and organized to follow a better adapted structure can also be established.

This decision is made according to the commitment to quality of the organization's responsible body. A statement of this commitment to quality needs to be included in the quality manual (see § 4.2). The wording should be reviewed by a person responsible for coordinating and carrying out successfully the establishment of the quality manual and for managing its ensuing development. Depending on the size of the organization, this quality representative (or "quality assurance representative) can also be the head of this organization.

After the manual has been drafted and before it is put into application, it must be signed by the director of the organization, thus committing him or her to its application.

3.2 Writing tips

Team work

The drafting of the different chapters and paragraphs of the quality manual is the result of a collective effort in which the various concerned departments of the organization are required to participate; their superiors have to be deeply involved in the work of drafting the quality manual, as they are themselves committed to putting into action the written provisions that concern them. Depending on the size of the organization, the drafting may be coordinated and supervised by a drafting committee or, if one has been formed, by a quality committee (see § 5.2.2).

An exact picture of the organization, not a "paper false front"

The quality manual must describe the true facts about the organization and the quality management provisions that are to be made. It is important that the work of preparing the quality manual evolve directly from the persons concerned and should in no way become a "paper false front" or a "cosmetic" measure to offer a flattering response to external requirements; therefore, it must never be managed by persons having little knowledge of the organization's activity or having roles of little responsibility role within this activity.

With this in mind, it is absolutely essential to avoid simply applying a typical ready-made ("ready-to-wear") manual such as a software package. If a consultant is called in, his or her action should involve only preliminary diagnosis, help with reflection and decision-making, and also, if this is the case, in auditing and advising on quality system improvements: the consultant should not act as a substitute for the organization's concerned authors!

Status and rationalization of the documentation: a source of progress

When the drafting of the manual begins, the status and the breakdown of the various existing documents that could affect quality should be reviewed and the documents rationalized in all the departments concerned. It will be necessary to sort through the existing material and discard any outdated or useless documentation. This implies taking stock of the existing documentation, which will need to be described as it is, even if malfunctions or a need for improvement results from the process. It will be necessary to make do with an initial version that has the virtue of being realistic, then revising it as quality within the organization is upgraded and progress is made.

A clear and concise drafting

To understand clearly an interpretation - in particular when it is by external auditors - the drafting of the manual must be clear and concise and hence refer to written procedures, organizational notes and to any other application documents for details. Elements that are likely to change should be avoided in the manual: for example, rather than list names of persons, the functions should be designated, and reference should be made to nominative organizational charts, which should be kept up to date. The coherence of the different sections of the quality manual requires special attention. Redundancies must be confined to the bare necessities of simple reminders.

Attachments and appendices should be restricted to any supplementary information (detailed organization charts, lists of procedures and instructions, examples of documents, etc.).

The case of multiple manuals

In the case where an organization has several departments or areas, a quality manual (and/or a quality assurance manual) can be drawn up (see chapter 2 and pyramid of figure 2.2) to concern:
- the overall organization;
- a single department, area or activity;
- one of the functions of the organization (design, production, etc.).

These manuals must be coherent with each other and must refer to the overall organization manual and, where applicable, to the general procedures concerning them. By applying a quality policy defined by general management, they stipulate the specific conditions under which this policy is to be applied in the area or function concerned and the authorities' corresponding commitments to quality.

Update

The quality manual will change in accordance with the setup and nature of the organization's activities and with the improvement of the quality system. The primary reasons for this change are:
- changes in the operational and functional setup;
- changes in the provisions madfe, in application procedures and documents, especially following corrective and preventive actions, quality audits, management reviews, etc.

The setup of the manual should facilitate its updating (see § 3.3.1 below).

An updating procedure (periodicity, generating factors of change, etc.) must be planned and mentioned in the section entitled "quality manual management" (see § 4.7).

3.3 Setup and structure

3.3.1 Setup

The setup of the quality manual must be designed to make its updating easy: classification and numbering of the different chapters and paragraphs, insertion of pages and stapling methods, etc. This setup and updating capability are particularly important in a quality assurance manual intended for commercial or contractual use or for the certification of the quality system.

The first page - or flyleaf - (of the manual and/or of a chapter)

This page must indicate any amendment (or revision) made, the date and the principal object of the amendment, the name of the person responsible (author or quality representative, with the signature of that person, where applicable), the name and signature of the verifier in charge or of the director (see figure 3.1).

If applicable, the mention "unrestricted release" (or "unrestricted" or sometimes also "uncontrolled") may be added. If a translation into a different language has been made, specify to what extent that translation is binding on the issuer of the manual.

C					
B	1.10.97	Update	F. Dupont	P. Thomas	
A	1.10.96	Rewrite after audit	J. Duval	P. Thomas	
O	1.03.95	Update	J. Duval	P. Thomas	
Index	Date	Name/Sign. Object	Name/Sign Written by	Verified by	Name/Sign. Management

Figure 3.1 Example of the first page of a manual
(or of a chapter)

Amendment (or revision) indication

This indication may be numerical or alphabetical.
The two best ways of replacing amended pages at a given date are to replace:
- all the pages of the manual or chapter with their first page (or flyleaf) so that they all bear the same document issue indication; or
- only the amended pages. In this case, a table stating each modified page and document issue indication together with the amendment date must be added (otherwise there is no verifiable record that the amended pages have been replaced). When the table relates to a chapter, it may appear on the first page of the chapter; when it relates to the whole manual, it is generally necessary to include the table on a second flyleaf).

Models of pages

Regardless of layout and design, practical features that every manual should contain include the folowing:
- organization name and logo,
- title of manual: quality manual, quality assurance manual, quality management Manual, Division A Quality Manual, etc. (see § 2.2.1),
- page numbering
- issue indication
- amendment date
- if applicable, signature of the persons concerned.

Page numbering

Rather than paginating the manual in one sequence from beginning to end, it is better to paginate by chapter:
- contents: page 0,
- chapter I: page I.1 to I.x,
- chapter II: page II.1 to II.y,

Consequently, the addition or deletion of pages will affect only the numbers of the pages of the chapter that is being modified.

Binding

Hardback or plastic covers are recommended. The binding system depends on the method used to replace the modified pages:
- if all the pages are replaced at the same time, a spiral or plastic binding is appropriate;
- otherwise, the binding system must allow for the pages to be replaced without having to use a special tool. The pages, however, should be sufficiently secured so that they will not readily come out of the binding.

3.3.2 Quality Manual Structure

A table of contents, particularly one that is clear and descriptive, is an important tool for drafting the quality manual. It can also be an invaluable help to the organization's internal users and to the auditors.

The structure of this plan can take various forms depending on the nature of the organization concerned (large medium-sized or small business, technical center, administration, etc.), or on its quality policy (quality assurance, quality management, total quality management, etc.). It may also depend on its place within the organization: "covering" manual (for the overall organization) or area manual (see figure 2.2). As no universal model is available, one of the following structures can be adapted.

Examples of structures

First part: introductory clauses

- Table of Contents
- Quality commitment statement by the head of the organization

- Scope and field of application
- Terminology and abbreviations
- References
- Presentation of the organization
- Management of the manual

Note: In order not to distract from the key elements, fold-out diagrams or unessential information should be included as attachments.

Some more detailed information on drafting is found in chapter 4.

2nd Part: Quality system (or any other title adapted to the content)

This part of the manual is essential for clarifying the provisions made to implement the quality policy. Depending on the nature of the organization and on the scope and field of application of the manual, the following might be adopted for the plan:

• *Option 1: Plan of the clauses of one of the ISO 9001, ISO 9002 or ISO 9003 standards*

This is the plan best suited for a "quality assurance manual," that is to say, one intended to meet a contractual requirement or to certify the quality system by taking one of these standards as reference. This approach is well designed for meeting the requirements of the standard concerned, and it facilitates the auditor's task without omitting anything.

An example of an organization dealing with cosmetics and applying the ISO 9003 standard can be found in figure 3.2.

The order of the sections does not need to be followed literally. For standards, there is not always an obvious logical order, as some subjects are discussed in several sections. Therefore, it is desirable to adjust or complete the plan to fit the practice of the organization (engineering,

SOPROCOS Management	SOPROCOS QUALITY MANUAL	QAM-01 Issue N° 4
Drafted by: Ph. Dequince	**SOPROCOS** **QUALITY MANUAL**	Page 1/28
Issued by: Q. Deucher		93/04/93
Validated by: R. Schneider, A. Rodier		

UNRESTRICTED
RELEASE

01 -Contents

Figure 3.2 Example of quality manual structure following the plan's sections of the ISO 9003 standard

service, development, transport, etc.) or the contractual or regulatory complementary requirements (maintenance, safety, etc.). An auditor must not refuse a plan different from the one used in the standard; but, to facilitate its reading, at the end of the manual, a **table of correspondence** showing cross-references between this manual and the relevant sections of the standard should be added.

However, this option 1 may prove to be restrictive if the quality management objectives need to be developed further: ISO 9004-1, etc.

• *Option 2: Plan of the sections of ISO 9004-1 standard*

When the quality policy requires a quality management approach based principally on the use of the guidelines provided in ISO 9004-1 standard, this plan may then facilitate the work of preparing a "quality manual" (or quality management manual). Under these conditions, if an organization already possesses this type of plan and wants to meet a contractual requirement or prepare for certification by using one of the three models mentioned above, the following can be applied:

- if the manual does not contain any confidential information and meets all the standard's requirements, present it as it is, with the addition of a **table of correspondence** showing the interaction of the respective sections as in the previous case; or,
- if the manual contains confidential elements, it is easier to extract these without changing the plan. However, as in the previous case, a table of correspondence showing the interaction should be added; or,
- if the quality manual is not a confidential document and if the existing quality system does not meet all the requirements of the targeted model, the introduction of complementary provisions and incorporation of an additional part to the quality manual entitled "quality assurance manual" will be necessary. The order of the sections contained in the standard in question will then be followed.

Figure 3.3 gives an example of this type of manual plan for an organization area manufacturing fibers and polymers.

Revision 01					Area : F and P	
Valid from:	Department	Service	Section	Level	Chapter	Page
17 june 91	00	QS	00	1	0-01	1/1

FIBERS AND POLYMERS AREA

Non-validated copy

CONTENTS

Presentation of the entity - assignments and responsibilities
   Presentation of the entity 4-01
   Area organization chart 4-02
   Assignments and responsibilities 4-03
   Managing Director's Statement and Commitment to 4-04
   Quality policy and objective 4-05
Structure of the quality system
   Structure 5-01
   Quality function assignments 5-02
   Organizational structure 5-03
   Documentation and records 5-04
   Audit and evaluation 5-05
Economic aspect 6-00
Quality in marketing 7-00
Quality in research and development 8-00
Quality in supplying 9-00
Quality in production control 10-00
Verification of products 11-00
Control of inspection, measuring and test equipment 12-00
Nonconformities 13-00
Corrective actions 14-00
Handling and activities following production 15-00
Training 16-00
Product safety 17-00
Use of statistical methods 18-00
Safety 19-00
Environment 20-00

Figure 3.3 Example of quality manual structure following the plan's sections of the ISO 9004-1 standard

Sotralentz	Sotralentz / Plastic department - F 67 Drulingen		
QUALITY MANUAL	Ref. QP- 01-001 - CHAPTER INDEX	Rev. 4	Signature QS :

Chapter modified by revision of the index

Chapter No.	Rev	Chapter title
3		
1	R1	Scope and field of application of the quality manual
Annex 1		Sotralentz quality charter, plastic department, of the 2.4.91
2	R0	Management of the quality manual
3	R0	Quality plans
4	R0	Definitions, terminology and abbreviations
5	R1	Plastic department organization
Annex 1		Sotralentz organization chart
Annex 2		Plastic department organization chart
Annex 3		Plastic quality organization chart
6	R0	Management reviews
7	R0	Plastic department quality policy
8	R0	Commercial production of regulated packaging
9	Pr	Authorization for mass marketing

	10	R1	Contract reviews
	11	R2	Purchase
	12	R0	Technical inspection for conformance
	13	R0	Control of nonconforming products
	14	R0	Control of critical elements
	15	R0	Corrective and preventive actions
	16	R0	Inspections and tests status
	17	R0	Justification of mass production
	18	R0	Initial mass product critical review
3	19	R2	Control of inspection, measuring and test equipment
4	20	R1	Control of quality documents and records
	21	R0	Production control
	22	R0	Training qualification
	23	R0	Handling, storage and packaging
	24	R0	Traceability
	25	R0	Products identification
	26	R0	Internal quality audits
	27	R0	Product qualification
	28	R0	Tests
	29	R0	Statistical techniques
3	30	R0	Quality system documentation

Note: Appendix 1 of index: "Manual / ISO-9002" table of correspondence

Figure 3.4 Example of quality manual structure based on sequence of operations

(*Source:* Sotralentz)

Main Parts	Detailed Sections	Corresponding chapters or paragraphs of this book (2nd part)
Introductory headings	Contents	
	Managing Director's statement of commitment to quality	
	Scope and field of application	
	Terminology and abbreviations	
	References	
	Presentation of the organization	
	Quality manual management	
	MANAGEMENT RESPONSIBILITIES	
	Quality policy	
	Organizational structure	
	Human resources	
	Personnel training	
	Management reviews	
	MANAGEMENT SYSTEM	
	Quality system	
	Quality planning	
	Document control	
	Traceability	
	Inspections and tests	
MANAGEMENT	Nonconformities, corrective and preventive actions	
	Quality improvement	

PROCESS

PRODUCT OR SERVICE QUALITY
Contract review
Design control
Design review
Purchasing
Production (or process)
Special processes
....

Life-cycle or "quality loop" phases

Installation and commissioning
Field operation (factory, equipment, transportation, etc.)
Introduction into the market, (sales offices, etc.)
Servicing (after-sales, maintenance, etc.)
End of the life-cycle (waste control, recycling)
RESULTS EVALUATION
Control of measuring equipment
Quality indicators
Quality related costs
Quality audits
Customer satisfaction appraisal

RESULTS

Figure 3.5 Example of detailed structure of quality manual distinguishing among management, process, and results

• *Option 3: Order of the logical sequence of activities*

This option offers the advantage of clarity of description when the organization's activities are such that they cannot easily follow the sections plan of the ISO standards mentioned previously. This also applies when the manual's field of application extends beyond the area covered by these standards, or when the manual pertains to a specific sector. This can be the case for:

- a services company,
- a laboratory,
- an administration,
- an organization wanting to group together in the same manual provisions pertaining to quality, operational safety or reliability, care of the environment, etc.,
- a sector or plant manual that is a sub-document of the applicable overall organization manual,

An example of this type of manual plan, also following the ISO 9002 standard and applied by an organization manufacturing polyethylene packaging for the transport of hazardous materials, appears in figure 3.4.

• *Option 4: Order for a specific quality system model*

This is an unconventional (or "non-standard") plan that can be applied when the quality policy is, for example, a quality management approach based on a conceptual diagram specific to the organization. In the example provided in figure 3.5, the quality system elements are represented by three distinct principal entities:

- **Management**: generic system elements (common to all the phases of the life-cycle).
- **Process**: system elements in the same order as the elements of the product's life-cycle (creation of a product); this order can be tailored to an organization's specific activity.
- **Results**: other generic system's elements.

Note that this structure is similar to the one adopted by the EFQM (European Foundation for Quality Management) for the French Quality Award (see Par. 16.5 and figure 16.3).

Whichever option is chosen, if the manual is presented in reference to a standard and if a section of that standard is not applied, a justification of the reason needs to be included in the manual, for an external auditor, for example.

Figure 3.5 can also be used as an introductory table to the content of Part Two of this book.

PART TWO - CONTENTS OF A QUALITY MANUAL

4 Introductory Pages and Annexed Documents

As seen in Chapter 3, the structure of a quality manual can take various forms. Examples have been provided where the introductory sections are grouped together to form the first part of the manual. Most of the essential content of this first part corresponds to the content of the X50-161 documentation sheet.

4.1 Table of contents

This section, which can serve as a working plan while preparing the manual, lists the various titles, numbered parts, chapters and sections, and the related appendices or tables.

The contents must form a clear and logical whole so that an auditor can easily find the principal elements of the quality system. To this end and where applicable, it can be completed by a table of correspondence showing the interaction (see: paragraph 3.3.2) between the sections of the manual and those of an applicable standard, or by showing access through a list of keywords.

4.2 President's statement of commitment to quality

The types of activities and the organizational structure of the organizations can be very diversified (service businesses, manufacturing concerrs, administrations, etc.), which gives scope for different types of statements.

The content of this statement expresses the decision and commitment to quality taken at the highest level of the organization:
- to define and write down the quality policy and objectives of the organization,
- to make this policy known, understood and implemented at all levels,
- to this end:
  . to ensure that the provisions contained in the quality manual translate this policy and these objectives in an effective manner, and are integrally complied with, within all the areas of the organization,
  . to ensure that the implementation and efficiency of the quality manual's provisions are assessed at all times.
  . to ensure the constant improvement of the quality system in accordance with the evolution of the quality objectives.

When the President decides to delegate the implementation of the quality system, the establishment and update of the quality manual, he or she should clarify this delegation in the declaration.

As an example, for a company:

*For continuity with all the previous actions taken by the company since its creation in 19..., and in order to answer the expectations of our customers with regard to prices, deadlines and **quality**, also in the context where, within a very competitive market, its partners (customers, authorities, public, shareholders, personnel) express their needs and requirements, I have decided to apply a **quality policy**. This policy aims at satisfying the customer at the lowest cost and focuses on the optimi-*

zation of resources, on the competence and motivation of the personnel, on good internal administration and on full compliance with legal obligations.

I commit myself to apply this policy, therefore I delegate the responsibility of implementing the system described in the current **quality manual,** *which will become operational from..........., to Mr. X, Quality Director (^{1}). He will be responsible for coordinating the quality objective by way of dialogue with all the departments concerned, for verifying that these objectives are reached, and for reporting his progress directly to me.*

Signed: [President]

Of course, this declaration can take various forms, depending on the nature of the quality policy and on the level of requirements. Also, where applicable, in the case where an organization is already covered by a general manual, it will depend on the situation of the organization within its overall structure. In this case, the declaration will reflect the commitment to quality already stated at a higher level of management. It is an integral part of the quality program and must be signed strictly by the head of the organization (or of the sector concerned).

4.3 Scope and field of application

Depending on the nature of the manual (quality manual, quality assurance manual, quality and safety manual, quality and environment manual, general or sector manual, etc.), under this section, the quality provisions taken by the organization - and possibly the related aspects of

---

^{1} *The designation of this person, his title and responsibilities may be covered by an organization memo that is separate from the manual and to which it refers (see § 5.2.2.)*

safety, environment, costs, deadlines - are described, whether the manual is intended for internal or contractual use. Its field of application is also clarified: organizational activities or areas concerned (products, services, other functions).

4.4 Definitions and abbreviations

This section defines the meaning of the words used specifically within the organization or area of activity, while clarifying the abbreviations and symbols appearing in the manual. Wording from standards documents should be used whenever possible.

In particular, this section may include a brief glossary listing the main quality-related words used in the quality manual and in the principal documents and the corresponding standardized sources to which reference is made.

4.5 References

This section is useful in the case of the following examples:
- Area manual: in this case, a reference should be made to the manual used at a higher level, from which it will form an application document.
- The quality policy is based on the application of a regulation, directives involving an industrial group or an administration, codes or standards, etc.: a reference should be made to these documents.
- An earlier situation needs to be mentioned: for exampole, the decision to alter policy or change documentation structure.

4.6 Presentation of the organization

In this section, it is desirable to give general and concise information about the organization, but without going into too much detail as this

issue will be dealt with in the section devoted to management responsibilities (see chapter 5):
- designation, legal status, capital and headquarters or place within
  an industrial or administrative structure, etc.;
- principal offices, addresses and means of communication (tel. fax,
  e-mail, Web site information);
- the organization's field of activity (products and services provided,
  technical and administrative functions, and possibly the techniques
  implemented);
- production, testing and inspection capacity;
- breakdown of labor force (percentage of managers, qualified spe-
  cialists, etc.);
- where applicable, reference to approvals secured from some customers (e.g. SIAR: Service de Surveillance Industrielle de
  l'Armement), or to the quality system certification (AFAQ "French
  Association for Quality Assurance," for example) and to certified
  products, etc.

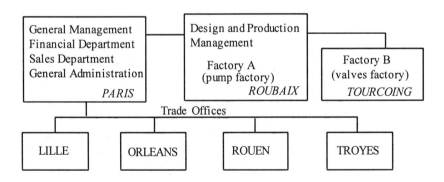

Figure 4.1 Example of organization layout diagram

Reference might also be made to an existing presentation document including a simplified diagram of the organization or establishment (see figure 4.1)

4.7 Control of the quality manual

This section clarifies the provisions taken for the design, the update, the distribution and the archiving of the quality manual or a part thereof. These provisions include:

- for creation or modifications, the appointing of a person in charge and the defining of the methods of deliberation and approval committing the various departments of the organization concerned (the part to be played by a drafting or quality committee);
- for distribution and archiving, the designation of an administrator responsible specifically for:
  . keeping records up to date (see § 3.3.1: if the first page of the manual and/or the chapter makes no mention of previous editions or amendments, then this information must be given on a separate page to show that previous editions have actually been distributed to the recipients).
  . verifying the release (often referred to as **"controlled,"** or **"managed" release**) (see § 7.4): in this case, each manual is given a number, allocated in accordance with an enclosed distribution list, with acknowledgment of receipt; the update is then ensured (unless specified, the quality manual remains, generally, the property of the issuing organization, which can ask for its destruction or return);
  . administering **unrestricted distribution**: in this case, the fact that updates are not ensured should be stated on the copy of the manual;
  . archiving: maintaining background evolution of the quality manual or of parts of it.

The scope of this section can be shortened by referring to an existing procedure.

4.8 Annexed supporting documents

The appendices should mention any unessential detailed documentation so that the main text is kept clear and concise, for example:
- reference documents, charters and group directives,
- tables of abbreviations and glossaries,
- detailed layout diagrams,
- lists of procedures and instructions,
- models of documents or forms, etc.

2 Top Management's Responsibilities

The title of this chapter might well be used for a chapter or section of the manual, especially if the structure follows that of one of the ISO 9000 standards (see § 3.3.2 options 1 or 2). Additionally, the following sections can be included as part of the manual's basic structure.

Contents of the quality manual

To put into effect the management's commitment to quality (see paragraph 4.2), the manual should clarify the corresponding policy and objectives and describe the setup implemented for these objectives to be reached, particularly with regard to:
- operational and functional structure,
- assignments and responsibilities,
- management of the technical resources,
- management of human resources,
- improvement of the quality system (this concept is dealt with in the ISO 9004-1 standard, but not in ISO 9001, 2 and 3).

5.1 Quality policy and objectives

Although the guidelines of the quality policy are already mentioned in the President's commitment to quality, this section should clarify some aspects with respect to:
- the stated and implied needs of the users of the products or service,
- the regulatory requirements (safety, environment management, etc.),

- the general policy requirements (profitability, place within the competition, etc.),
- unusual aspects of certain areas of the organization.

This policy generally implies objectives and deadlines, as indicted by the following examples:
- making personnel aware of quality and to train them,
- designating the persons responsible for setting up or improving the quality policy,
- extending the quality system to certain areas or functions,
- ensuring the qualification of some processes or equipment,
- setting up quality measurement indicators,
- training quality auditors,
- preparing the organization or one of its areas for the certification of the quality system, or of its products,
- preparing the presentation for a quality award,

The policy should describe how management needs to proceed to make this policy and these objectives known and understood throughout the organization. With regard to external auditors, existing documents that can be applied should be referred to as often as possible.

An area (or department) manual while outlining the specific quality objectives of the area concerned.can also be used to convey information about the general organization-wide policy.

5.2 Organizational structure

Management is responsible for establishing the organizational structure and the technical and human resources required, together with defining the assignments and responsibilities to show how they directly have an effect on achieving of quality. The quality manual must give a clear and precise image of this.

Note: In ISO 9001, 9002 and 9003 standards, the section entitled "Organization" appears in the chapter "Management responsibility," while in the 9004-1 standard, it is in the chapter entitled "Structure of the quality system"; this can be taken into account for choosing the plan that will be adopted in drafting the quality manual (see § 3.3.3).

5.2.1 Hierarchical and functional organization chart

This organization chart will have to be more complete than the one that will eventually figure in the organization outline (see figure 4.1); it must permit the organization and operation of individual functions or processes to be understood. It is advisable to distinguish clearly the **hierarchical links** (administrative authority over the personnel and their activity) - which might be depicted by a straight line, - and the **functional links** (non-hierarchical links that correspond to well-defined assignments) - depicted by a dotted line, for example, for non-hierarchical quality functions - (see figure 6.3). Finally, where applicable, the organization's relationship with exterior bodies governing quality issues should be indicated: **interface link relationships** with other areas of the organization covered by a specific manual, or with an inspection or maintenance service outside of the organization, with a partner, with an administration, etc.

To be consistent with this organizational chart, a succinct definition of the general attributions of the different units involved with quality should be stated. The assignments, responsibilities and authority of the persons in charge of managing, performing or verifying the work affecting quality should also be stated in sufficient detail. In this case, reference is made to production or servicing tasks that are essential as well as those concerning **interfacing between** the organization's **internal and external departments**. The quality function (see § 5.2.2) should be described elsewhere in a more detailed manner.

In the text of the manual and in the organizational chart, it is advisable to refer to personnel in terms of their functions to make it easier to keep the manual up to date. It is helpful to have a nominative organizational chart, which can be updated and distributed separately amd which can be used by auditors. To convey a better understanding of the organization, it is sometimes useful to include the number of staff and their allocations.

5.2.2 Choice of the quality function

It is one of management's major responsibilities to pass on the quality policy and objectives to the various appropriate hierarchical levels and ensure that they are implemented, followed and arbitrated by a quality structure, i.e. one or several appropriate lines of communication between the President to key members of the organization. Depending on the terms of the President's commitment to quality (see § 4.2), these levels might consist of:

- a Quality Committee,
- a Quality Department or a Quality Assurance Department,
- a "Quality Vice-President or Director" or "Quality Manager" or a "Quality Management Representative in Charge of Quality,"
- a "Quality Assurance Director" or "Quality Assurance Manager,"
- a network of quality coordinators or correspondents, or
- a combination of the above, especially for large, diversified organizations.

The Quality Committee

• Background

In many small, medium-sized and large organizations, it is often from top management's decision to create a quality committee that an organization takes its first steps towards implementing the structures needed to ensure manufacturing or service quality.

After working for several weeks or months, for one or two hours a week, this committee is often able to put propositions to the top management, such as:

- the necessary structure: personnel, financial resources;
- objectives, a program and deadlines;
- a series of written procedures;
- improved technical resources: measuring, etc.

• **Composition**

This varies enormously, depending on the size of the organization:
- for small plants, the plant manager and his or her secretary, the workshop manager and the quality control manager, could form a highly operational committee.
- for larger organizations, say, of around 500 people, a Quality Committee might include:
  . the Managing Director or the person delegated to set up the quality system (see § 4.2); this person would chair the quality committee,
  . persons representing services or departments involved in quality: quality or quality control, technical, methods, purchasing, sales and customer service departments,
- for a large organization having multiple divisions or establishments (factories, research centers, warehouses, sales outlets, etc.), there could be:
  . a Quality Management Committee at the highest level of the organization, with the Managing Director, top executives and the Quality Manager.
  . quality committees within the sectors or establishments, with the Quality Director or deputy and the operational people in charge of the divisions or establishments.

Though it may sometimes be necessary to call emergency meetings, these committee meetings, lasting one or two hours, typically can be held on a monthly basis. The committee chairman can define the agenda. The minutes should be distributed to all the committee members and to the relevant authority (managing director or quality manager, depending on the case).

• **Quality committee assignment**

Priority objective:
To resolve major quality problems brought to the attention of the quality manager during the month preceding the meeting (in the case of a monthly meeting).

Permanent assignment:
- Initially, to propose a quality policy to the managing director and then ensure its implementation;
- Depending on this policy, to define the long- and medium-term quality objectives;
- To supervise the evolution of techniques and methods according to objectives and circumstances;
- Approval of the quality plans:
  . strategic quality plans (see § 2.2.2)
  . quality plans relating to important or new supplies;
- Followup to ensure that the quality objectives are met;
- Arbitration: settling of conflicts between departments or difficulties that have not been resolved through the usual channels.

The Quality Department

Within organizations that are big enough and with the necessary re-sources, a General Manager may decide to set up a department headed by a Quality Director or Manager supported by a small team in charge of implementing the quality policy and objectives. Depending on the chosen quality policy and the ensuing quality system (see § 6.2) this management can:
- serve simply as a non-hierarchical entity: for example, to manage the quality policy, without any authority over operational quality control functions,

- have one or several operational departments under its authority: quality department, control laboratory and, in certain cases, safety and environmental protection department, etc. (see figure 6.3).

• **The "Quality Director" or "Quality Manager" or "Quality Management Representative"**

This is one of the titles often attributed to the person appointed by the Managing Director in his or her commitment (see § 4.2) to implement the quality system, when the quality policy is a quality management policy covering certain aspects other than quality control and quality assurance (for example, those of the ISO 9004-1 standard).

• **The "Quality Assurance Director" or "Quality Assurance Manager"**

These titles are often used when the quality policy is confined to product or service quality control and quality assurance aspects that are contractual or regulated (for example, in order to apply one of the ISO 9001/9002 or 9003 standards).

The choice of the preceding titles and functions depends on the size of the organization, its hierarchical traditions, the responsibilities delegated by the Managing Director and on whether the function is carried out full time or part time.

The function can be carried out part time as long as it derives directly from the top management of the organization or plant concerned and is carried out independently with respect to the normal hierarchy of operations.

• **The quality "coordinators" or "correspondents"**

Within an organization or one of its divisions, when so justified by the size, these persons form a network appointed within the various plants of the division and report directly to the plant manager. They are generally part-time personnel in charge of quality or quality assurance within their own plant and represent their respective managers to the quality or quality assurance director or manager at the higher level (see figure 6.3).

In light of the above, the person designated by the Managing Director should define, as part of the quality system, the assignments and responsibilities of those who have a quality function. Examples of these functions are given in paragraph 6.2 (quality function).

5.3 Human Resources

The identification of the training needs and staff qualification and of the provisions to be made to meet quality objectives are the responsibility of management. Management is also expected to motivate the personnel within the quality management approach inspired, for example, by the guidelines of the ISO 9004-1 standard.

Depending on the drafting structure being used, these provisions can be described:
- as part of the "Management responsibilities" and "Training" sections (structure of 9001, 2, or 3 standards), or
- in the "Quality system" chapter of the "Resources and Personnel" section or of the "Personnel" chapter (structure of the 9004-1 standard), or
- in the description of the human resources management or personnel department.

This subject will also be dealt with in paragraph 6.2 (quality function) and in chapter 15 (training).

5.4 Management reviews

The management review is a "formal evaluation by top management of the status and adequacy of the quality to quality policy and objectives" (see ISO 8402, paragraph 3.9).

This is an important management responsibility, because it is from such reviews that management becomes aware of quality policy results and of possible deviations. It can then decide upon general corrective action or can reorient its quality policy and objectives, using resources accordingly.

A management review is prepared by the quality manager or representative using the quality results collected within the quality departments, from the people involved in quality (quality correspondents, etc.). The review systematically takes into account the internal or external quality audit results (those of the customers or authorities, where applicable) and management evaluates the efficiency of the ensuing corrective action.

Management reviews also include, where applicable, analysis of existing guidelines and results of customer satisfaction surveys. When a Quality Committee has been established, it takes part in their preparation and progress.

For the quality system to be efficient and to meet the requirements of contracted and audited standards, it is therefore important that such reviews be scheduled at suitable intervals (at least once, but preferably twice, a year). **Reports** of the review **must be preserved**, specifying the agreed actions, the persons responsible for them and the target dates. These reports must be suitable for submittal to a quality auditor.

The fundamental modifications of the quality manual often depends on the decisions made during such reviews.

6 The Quality System

The efficiency of the quality system depends essentially on:

- the identification of the activities within the "quality loop";
- the efficiency of the organizational structure for dealing with each activity and the related procedures, processes and resources, and on the corresponding documentary system.

Contents of the quality manual

In the manual's outline, the provisions made to implement the quality policy can be grouped together in a section entitled "Quality system" (see § 3.3.2). This title may cover only one section or heading from the manual, if the plan of the ISO 9000 standards series is followed. However, in this case, there is a problem: the ISO 9001, 9002 and 9003 outlines differ from that of the ISO 9004-1 standard and therefore a choice must be made between one of the two types.

This chapter clarifies for the authors of the quality manual the different aspects of the quality system and is designed to help them "construct" their own system as a function of the activity concerned and the defined needs. It should also provide guidelines on the system elements that need to be described in the manual.

These elements can be allocated to appropriate sections of the manual outline or plan adopted.

6.1 The quality loop

Quality standards must be applied to all the activities concerned with the corresponding quality policy and objectives of the organization. The quality manual therefore must, in the first place, stipulate what these activities are.

In all cases, all phases ranging from the stated needs of the customer through to the customer's ultimate satisfaction should be identified.

When a product is being supplied: "life-cycle" phases

Refer to the "quality loop" diagram (see figure 6.1).

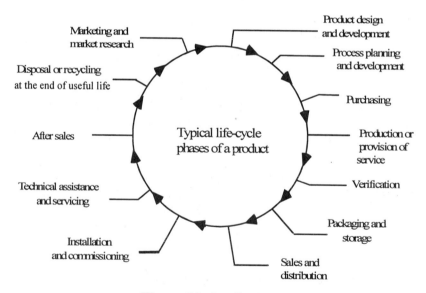

Figure 6.1 Quality loop
(*Source:* ISO 9004-1)

The activity concerned should be broken down into successive phases: examination of requirements, planning, design (where applicable), preparation for production, etc., and for each identified activity, the following should be described:

- How the activity is **controlled**:
  - . the resources and personnel required to accomplish the activity, and any relevant qualifications,
  - . the responsibilities of those involved, especially for quality verification;
- How to obtain **confidence in this control**: planned quality assurance provisions and role of quality function to verify its correct application.

It is the **quality plan** specific to this supply (see § 8.3) - that gives a synthesis of these successive phases and designates the procedures to be applied for each phase - and it is these procedures that clarify the **quality control and quality assurance actions,** which are often interrelated (see "Horizontal arrows" of figure 1.1).

Special case of a sub-contracted supply

The above remains applicable, but the quality loop can be reduced, for example, to purchasing, production and verification (inspection). The quality plan can be an extract from the customer's quality plan (see left part of Figure 2.2) or can take the form of a simple manufacturing production sheet (see § 11.2).

Case of supply other than a material product

If the activity consists of providing a service (for example, operation of an existing installation, maintenance, administration, etc.), the quality loop must be clearly identified. For a service supply, the ISO 9004-2 standard "Guidelines for services" may apply and the "service quality loop" could be a guideline (see figure 6.2).

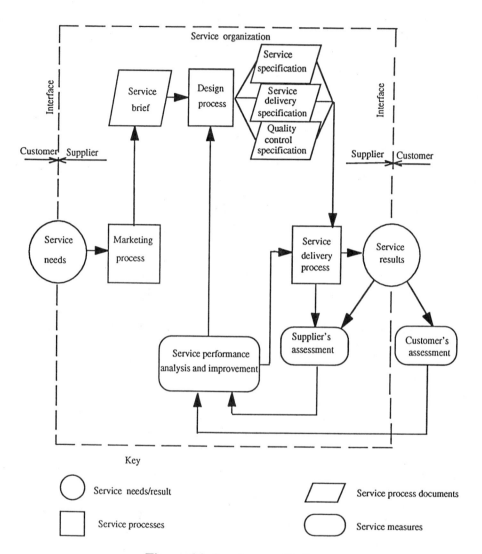

Figure 6.2 Service quality loop

In every case

For all these activities, **necessary and sufficient provisions** should be made so that the quality system thus formed **establishes confidence** that the set quality goals will be fulfilled with the desired efficiency and at the lowest cost.

6.2 Structure of the quality function

The quality function is the activity of people appointed to apply the quality policy and to ensure that the quality objectives are met (see 5.2.2). They are the persons in charge of verifying activities:

- in the **operational line**, they are persons in charge of verifying the **effective quality control** of the products or services from the design phase through to the final satisfaction of the customer: they verify the design, pilot the special processes during the production phase, (welding, etc.), carry out inspections and tests, etc.
- in the **functional line**, independent from the operational line, they are in charge of **establishing confidence** that quality requirements will be met: they participate in design reviews, supervise inspections, and carry out quality audits.

To describe the structure of the quality function involves:
- describing the **authorities and functional link relationships** existing among all those persons involved in quality (see § 5.2.2): Quality Director or Quality Manager (or Quality Assurance Director or Quality Manager), coordinators, inspectors, supervisors (surveillance), etc.;
- detailing the principal **responsibilities and assignments** of the aforementioned persons by specifying their authority to avoid the appearance of nonconformities and implement the necessary corrective decisions.

Forms of quality structure

The quality function structure can take various forms, depending on the nature of the organization and on its quality policy. It depends principally on:

- the nature and frequency of the quality control: importance and role of self-inspection, inspection allocated within a specific or centralized production plant - in this case coming under general management, production or quality management,
- the choice between the "quality" functional line (quality system + product quality) and the "quality assurance" functional line (quality system only),
- the existence or not of a Quality Department or Division or of a Quality-Safety Department or Quality-Safety-Environmental Department.

Figure 6.3 gives some examples.

Examples of quality assignments

The assignments and responsibilities depend here again on the choice between "quality" (in the broad sense of management) and "quality assurance."

For a Quality Assurance Manager:
- Ensuring that the procedures and documents enabling the application of the quality assurance manual (or quality manual) are established and kept up to date.
- Ensuring a quality assurance advisory and information link among production plants and related processes.
- Verifying or that the corresponding activities are carried out in accordance with the defined requirements, particularly in accordance with internal quality audits.
- Identifying activities requiring corrective action and ensuring their implementation.

- Establishing an internal audits program and ensuring its implementation.
- Organizing suppliers' evaluation and surveillance with regard to quality assurance.
- Representing the organization externally with regard to quality.
- Participating in quality training and information actions.
- Reporting to management on the actions taken and to propose improvements to the quality system.

For a Coordinator or Quality Assurance Correspondent:
- Participating in the establishment of quality procedures and documents.
- Assisting the plant manager with regard to quality assurance, principally with activities concerning:
  . the supervision of compliance with applicable documents,
  . the management of documentation,
  . the definition, where applicable, of the suppliers' quality assurance requirements,
  . the participation in "quality meetings" organized within the plant level or at a higher level of management.

For the quality system to be efficient, this function, which may need to be only a part-time activity, should be entrusted to one of the plant manager's direct assistants.

For a Quality Manager or Quality Coordinator - these positions usually-have a broader scope of functions - the assignments are of a similar nature, but assignments and responsibilities relating to the technical quality of products and services (technical surveillance, qualification of the processes, etc.) or assignments within the quality management may be added: implementation and followup of quality indicators, participation in quality improvement groups, etc.

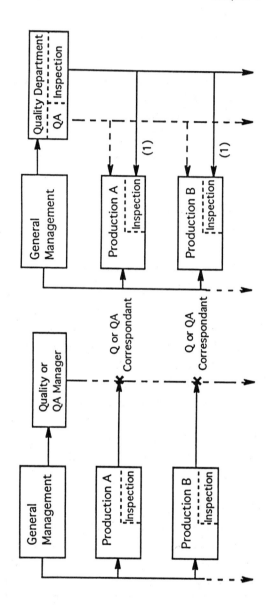

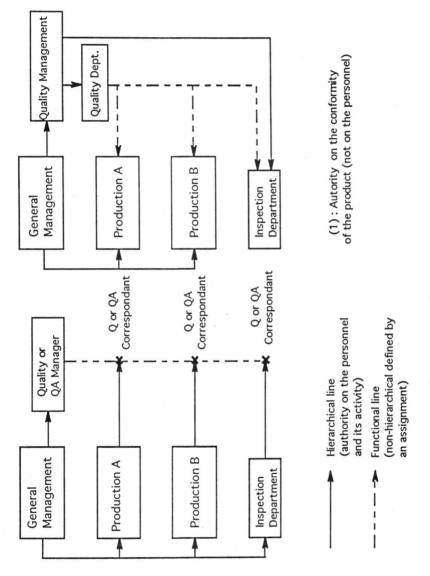

Figure 6.3 Examples of quality function structures

Inspection and surveillance function

This function that is often called "quality inspection" and that is more a "conformity inspection" tends to have fewer personnel, as the quality assurance provisions are implemented and as confidence grows within the organization (internal quality audits) or among suppliers. However, it is important to identify the points of those activities requiring conformity inspection and applicable cases where surveillance of this inspection needs to be scheduled.

For a product, project or particular contract, it is often necessary to specify the inspection and surveillance points in the quality plan (see § 8.4.2).

Indicate the units or the persons in charge of this function, their positions within the hierarchical and functional organizational chart, their assignments and responsibilities; to do this, reference could be made to the procedures or organizational directives.

6.3 Document structure

The different types of documents that contribute to the implementation of the quality system and that form the pyramid of quality documents (see chapter 2 and figure 2.1) should be described briefly. A distinction might be made between, for example,:
- **Organizational documents**:
   . a quality manual and, where applicable, several area quality manuals,
   . organizational procedures and organizational directives with associated appointment papers;
- **Quality plans, audit programs**, such as corrective or improvement action programs;
- **Defining documents**, such as those that describe the resources used, their operation, their qualification, etc.;
- **Technical documents**:

. translating requirements: standards, specifications, requirements files, etc.

. design and development files;

- **Manufacturing and operating documents**;
    - . operational procedures, operating methods (which define the detailed succession of actions necessary to carry out an operation) or processing sheets (which define the elementary operations necessary for the product to be made),
    - . orders (which determine rules that have to be complied with, especially concerning safety matters), working instructions, etc.;
- **Records or reports**, which furnish evidence that the activities, actions and verifications have been carried out and which keep track of the results obtained;
- **Interface documents** with customers or external organizations.

Increasingly, documents are in the form of computerized records.

To obtain an efficient but economical quality system, it is necessary to aim at:
- **what is necessary**: useful and rigorous documents required for the planned activities,
- **what is sufficient**: clear and concise documents that avoid useless paperwork.

For each identified activity, the necessary documentation must be established and kept up to date in order, in accordance with the principles outlined in figure 1.2:
- to plan and describe what needs to be done,
- to prove that work has been carried out in accordance with the plan,
- to keep records of what has been done,
- to provide confidence in the application of the preceding actions.

6.4 Control of resources and personnel

This is a fundamental principle of any quality system. This control must be ensured for all the identified activities of the quality loop: for the design, production, verification resources, testing, measuring and recording resources and for the different qualifications that may be necessary to ensure that quality is obtained.

This relies on:
- the availability of equipment, processes and procedures necessary to obtain conformance with the objectives.
- the designation of persons trained for the activities concerned,
- the definition of the corresponding qualifications that need to be formalized.

Content of the quality manual

The primary corresponding provisions may be described within the appropriate chapters or sections of the manual, and the various procedures may be used for describing the provisions to be applied in the field.

6.5 Contract reviews

The contract reviews are "systematic activities carried out by the **supplier** before signing the contract - and repeated at different stages of the contract - to ensure that requirements for quality are adequately defined, free from ambiguity, documented and can be realized by the supplier" (see ISO 8402, paragraph 3.10).

The term **"supplier"** is understood to refer to the entity that has to meet contractual requirements, i.e. the organization that has to be able to meet, dependably, the customer's requirements. This element of the quality system therefore directly concerns this specific type of organization.

That is why, within the ISO 9004-1 standard, for example, i.e. within the internal aspects of the quality management, this element is not directly mentioned; however, the ability to meet the needs required for a product or service is dealt with in this standard under the heading "Quality in marketing."

On the other hand, for contractual situations, it is an important element of the system because the customer needs to be assured that:
- the supplier has not "finessed" by promising to do more than he could to secure the order;
- the supplier has made all the required provisions to meet the requirements, especially if the actual order differs from the original tender or if, when the tender was submitted, all the necessary resources had not yet been implemented;
- during the contract, details affecte by modifications made to the contract or additional clauses have been passed on to the functions concerned.

Content of the quality manual

If the quality manual is to meet the requirements of the ISO 9001, 2 or 3 standards, one of the sections must refer to the contract review procedures and to the procedures for coordinating these activities. **Records of these reviews must be kept**, preferably as reports, and should be available for submission to a quality auditor.

6.6 Traceability

Once more, this is a very important principle for every quality system. Traceability is the "ability to trace the history, application or location of an **entity** by means of recorded identifications" (see ISO 8402).

In the ISO 9000 standards, the **entity** refers essentially to a material product; the requirements concern principally:
- **identification** of the materials and parts throughout the production process,

-   transfer of this identification to all the corresponding documents,
-   **recordings** that make it possible to find the history of activities or processes throughout the quality loop,
-   **archiving**: filing after use, for a determined period of time.

When applying these standards, the **entity** may refer to various product categories such as a service delivery, a software package, etc., or to a technical action such as calibration (reference to a primary standard or physical constant).

The traceability provisions required may vary greatly, depending on the activity of the organization, contractual requirements (which can lead to high costs) or the internal good management practice requirements.

These are described in different sections of the manual, corresponding to the activities described.

Traceability for a supply

This is a way of:
-   During manufacture, tracing the cause of a nonconformity or defect,
-   During use:
    -   following up the supply to confirm the expected life limits and to organize maintenance,
    -   to trace the origin of the raw material and manufacturing stages so that in the event of faults or defects, the cause can be identified.

The procedures should therefore, for all supplies concerned, guarantee traceability and identification of components coming from suppliers.

Identification

The identification of the products must be ensured (from the original components through to the delivery stage) as well as for documents (see

§ 7.5). It must be the object of procedures established in relation to the activity concerned.

Essentially, for a product, identification aims at:
- avoiding confusion during manufacture and ensuring traceability,
- avoiding the use of non-conforming or defective elements,
- identifying the corresponding documents.

It is useful to refer to the recommendations of the ISO 9000-2 standard (see § 4.8), regarding the various forms of identification.

Records and archiving

The records relating to quality give direct and indirect evidence that the product or service meets the technical requirements and is in accordance with the contractual and regulatory requirements. The corresponding provisions should be the object of procedures established in relation to the activity concerned.

The records can, depending on the kind of activity and on technical development, take the form of a hard-copy or microfilm, or increasingly may be computerized. The records concern nearly all the phases of the quality loop, and the methods used should be mentioned in the corresponding sections of the manual and procedures.

Archiving essentially concerns documents. But there are cases where models, samples, photographs etc. also need to be filed away.

What counts in terms of quality is:
- the quality of the supporting material,
- the security of the archiving location and the accureeate duplication of the supporting material,
- the minimum preservation time, which may be stipulated by contractual or regulatory requirements (see ISO 9000-2 paragraph 4.16).

Content of the quality manual

If the quality manual has to meet any of the requirements of one of standards 9001, 2 or 3, one of its sections must refer to the identification, classification, filing of documents as well as the procedures for eliminating obsolete documents (see § 4.5 and 4.16 of these standards) or to document management and control procedures (see § 7.2). Reference may be made to a **list of preserved recordings** (with defined duration), which are often specified by these procedures.

7 Document Control

7.1 Principles

The word "document" is used here to refer to information that has been stored in a medium (hard copy, tracing paper, microfilm, photograph) or that has been computerized.

It is impossible to establish an efficient quality system unless, wherever an activity influencing quality is taking place, the "applicable" or "valid" documents (correct, up to date and, where applicable, formally "validated": see § 7.3) are available at every utilization level. This is something that auditors are bound to check. This aim can be reached only at the lowest cost if the documentation management system has been adapted to the activities concerned has been. Ideally, this management is computerized.

It is therefore important to keep up to date - and to be able to submit to a quality auditor - a **list of applicable documents** for all activities influencing quality. Within an organization of a certain size, the lists of documents of several areas can be grouped together within a "folder of applicable documents" by establishing a methodical tree chart (often called "documents breakdown folder"). This list can be reduced to a "list of applicable procedures": each procedure then refers to documents applicable to the procedure.

Content of the quality manual

In a chapter entitled "Control of documents," which refers to various procedures, the system that is going to be used should enable people to:
- choose, from the documents concerned with the activity, those affecting quality;
- draft, identify and revise the documents;
- manage their implementation and circulation;
- ensure their archiving.

The manual must, more specifically, clarify provisions taken to establish and control the procedures.

7.2 Nature and selection of documents to be controlled

When a quality system needs to be implemented and a quality manual needs to be drafted, it is necessary to start by making an inventory of the existing documents and, **evaluating their importance with regard to quality,** to establish a "pyramid of quality documents" (see figure 2.1) based on applicable field documents.

At the level of middle and lower management, such documents consist of organizational memos and various directives issued essentially as general procedures. In departments and workshops, the documents used have to be sorted and formalized - useless or out-of-date ones should be eliminated - in order to establish the area procedures and various operational documents of the quality documentation structure (see § 6.3).

In an industrial organization, no attempt should be made at the "control" of or even "the putting into quality assurance form" of all the existing applicable documents, because this would overload the system. "Quality documents" should be sorted by type (manuals, procedures, instructions, quality plans, specifications, processing sheets, drawings, etc. (see § 6.3)).

The creation of customer/supplier quality interface documents and the management of the corresponding mail are options that should also be considered.

In an administrative organization, the selection of documents to be "QA-formalized" should be evaluated according to the quality manual and procedures. Management will decide the rest.

The results of this organizational work should be taken into account in a **procedure for document control and management**, established in the organization or in the different areas or departments to which the quality manual must refer.

Figure 7.1 shows an outline of such a procedure.

7.3 Creation and revision of the quality documents

In general, the following terms are used:
- **"Written by"** or **"Author,"** or **"Drafted by"**: the person in charge of writing a quality document;
- **"Verified by"**: the name of any person responsible for checking the document quality;
- **"Issued by"** or **"Approved by"**: the name of the department or function authorized to put the documents into circulation so that they can be used.

These actions are confirmed by the signing of these documents.

When the "quality assurance" function is involved, some of the organizations sign the document in the following manner:
- **"QA verification,"** before signing for "Issued by" or "Approved by," or
- **"QA approval,"** after signing for "Issued by" (this may have the drawback of requiring approval of a hierarchical superior by a subordinate function).

COGEMA

810 PR 007
RevOO 2/13

General Management
Quality (Head Office)

Procedure

Control of a plant applicable documentation

Table of contents

5. Identification of applicable documents

6. Implementation and circulation of applicable documents

7. Administration and breakdown of applicable documents

8. Modification

9. Cancellation

10. Archiving

Appendix 1: List of identifiers denoting COGEMA plants
Appendix 2: Notice of application (for information)
Appendix 3: Tree chart

Published on 18.02.93	
Drafted by	Verified by
Classification : O810	Classification plan

COMPAGNIE GENERALE DES MATIERES NUCLEAIRES

Figure 7.1 Example of a procedure for document administration and control

For the quality system to be efficient and to prevent it from being unwieldy, **signatures should be confined to what is strictly "necessary and sufficient."** For some documents, it might be useful to include only the signatures of two people involved in the task:

- for the person who has carried out the task or dealt with the subject: sign "drafted by" or "written by;"
- for the person hierarchically responsible, who has checked or approved the task: sign "verified by" or "issued by" or "approved."

On the other hand, four signatures may be necessary on some documents, depending on the scope of the quality actions concerned - for example, regardoing important matters of safety. The danger is that a plethora of signatures will become so much red tape, particularly since some documents are hastily signed (as a matter of form) and the signature adds nothing to the quality of the document.

Figure 7.2 gives examples of how signatures might be implemented.

The adopted provisions must be the object of one or several written procedures, to which sufficient reference should be made in the quality manual.

The documents should be revised and confirmed in the same way. The document should indicate the type of change, within the main text, or in the appropriate appendix. Any amendments should be indicated by a line to that effect in the margin.

7.4 Implementation and circulation of quality documents

The document implementation is the responsibility of the issuer. This person is the "owner of the document." He or she chooses the author and the person(s) in charge of verification and is responsible for the content and circulation but not the ensuing application, if it depends on another authority, as in the following.

Validation

If a document, in its applicable state, is likely to be used for several contracts, it may be **validated** for a given contract by the person in charge of that contract; this "validation" (see ISO 8402) means that the requirements specified for this contract are met by this document; it gives the "go-ahead" for its use and can be formalized with a complementary mention such as "Approved for implementation" or "Approved for application," and should bear the corresponding signature.

Distribution management

Distribution is managed by the department issuing the document or by the archiving department. It can be specified on the document itself or carried out by means of a mailing service, distribution or implementation note which carries the signature of the authority concerned. The addressees are specified either on the document itself, or via an attached distribution note or list.

A distinction is made between:
- **controlled release** (often referred to as "managed"): in this case each document is distributed with an acknowledgment of receipt and its update is ensured; this distribution is appropriate when there is a need to ensure that the users have the updated documents at their disposal;
- **uncontrolled release** is generally sufficient for documents distributed for information only.

Distribution of the quality manual and procedures

It is recommended that distribution be carried out as "controlled release" to the assigned people by the Quality Manager using an up-to-date list. The recipients must similarly distribute these documents to their departments or workshops and ensure that only up-to-date documents are used and that outdated documents are withdrawn.

Drafted by (1)	Verified by	
Drafted by (1)	Approved by	
Author (1)	Manager	
Inspector	Q. Manager	
Techn. Dept	Q. Department	
Drafted by (1)	Verified by	Emission
Drafted by (1)	Verified by	Approved by

Drafted by (1)	Verified by	Application

Drafted by (1)	Issued by	Validation

Established by	Verified by	Validated by

Techn. Dept.	Q. Department	Management

Drafted by (1)	Verified by	Verified by QA	Approved by

Drafted by (1)	Verified by	Issued by	Approv. by QA

Drafted by	Verified by	Issued by (2)	Validated by (3)

(1) "Author" or "Written by" or "Drafted by"
(2) Document generally valid
(3) Application specific to a contract

Figure 7.2 Control of documents: Examples of presentation of signatures

The complete set of manuals and procedures issued with the successive indexes should be preserved for an archival period (to be defined) by the Quality Manager.

7.5 Presentation and identification of documents

All the documents affecting quality must be carefully identified, numbered, indexed and paginated.

It is advisable to use the same formalized layout for each type of document and mention systematically:
- the type of document, (with a specific layout),
- its title,
- the issuing plant,
- a chronological identification number,
- an amendment date and index.

This information should appear on every page of an important quality document. Furthermore, on one of the first pages, signatures and any necessary identifications concerning amendments and application conditions, should be appear.

7.6 Preparation and control of procedures

Refer to paragraph 2.3 for the procedure definition, place and role within a series of quality documents.

The procedures ("written" or "documented" following the ISO 9000 standards) form the basic documentation for the implementation of the quality system, i.e. for the planning and control of quality-related activities. They must explain, with the **necessary and sufficient amount of detail**, the way in which the different activities are to be carried out. They may be complemented, where applicable and especially for added technical detail, by documents such as: work instructions, operating methods, etc., and to which they refer.

As with the quality manual, (see § 3.2) the procedures should be a team effort and should be clear and concise (the ideal procedure should require only one or two pages!). For technical details, the procedures should refer to work instructions, operating methods and other detailed operational documents.

An efficient way of starting to establish a procedure is a "flow diagram" summarizing the successive actions and decisions needed. This flow diagram should be attached to the procedures and will facilitate its field application. Figure 7.3 illustrates an example of such a diagram for the inspection of incoming goods in an agricultural cooperative.

Generally, a procedure should be prefaced with an explanation of how it should be established, approved, identified, distributed and kept up to date, in order that all the ensuing procedures form a coherent and rigorous system to cover all activities (This introduction is often called the "procedure of procedures.")

A list of applicable procedures (with amendment dates and indexes) should be kept up to date. An extract (without date or index) may be attached to the quality manual.

Figure 7.4 is an example of a list of procedures attached to the manual.

7.7 Technical documents control

In some organizations that have a specific need to control a great number of technical design and manufacturing documents, a centralized organizational structure for the management of technical documentation should be implemented. Following are some examples of provisions taken to this end.

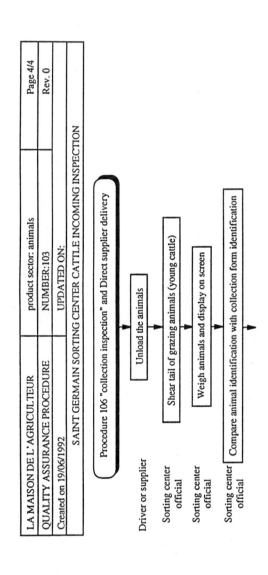

LA MAISON DE L'AGRICULTEUR	product sector: animals	Page 4/4
QUALITY ASSURANCE PROCEDURE	NUMBER:103	Rev. 0
Created on 19/06/1992	UPDATED ON:	
SAINT GERMAIN SORTING CENTER CATTLE INCOMING INSPECTION		

Procedure 106 "collection inspection" and Direct supplier delivery

Driver or supplier — Unload the animals

Sorting center official — Shear tail of grazing animals (young cattle)

Sorting center official — Weigh animals and display on screen

Sorting center official — Compare animal identification with collection form identification

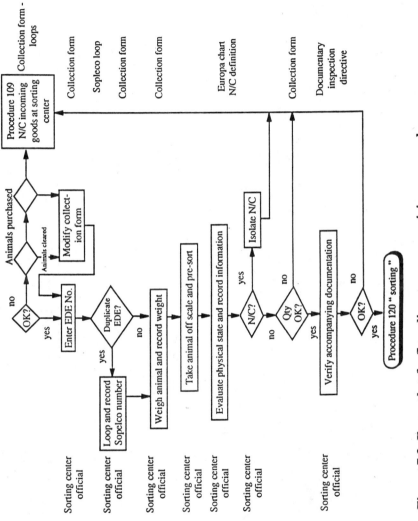

Figure 7.3 Example of a flow diagram summarizing a procedure
(*Source:* FRCAL)

Quality
Assurance
Legrand

QUALITY MANUAL

LIMOUSIN PLANT QUALITY SYSTEM

§ ISO 9002 STANDARD	Procedures and common instructions	Specific procedures and instructions by technology			
		Reinforced plastics	Conversion of plastics	Conversion of metals	Assembly
4.1 Management responsibility	AJI028				
4.3 Contract review	AJP060 AJPOO4-AJP006 AJI005 AJI061-AJIP062 AJI040				
4.4 Document control	AJP 008 AJI009-AJI041 AJI049-AJI062 AJP001-AJP018 AJP019				
4.5 Purchasing	AJP031-AJP032 AJP033-AJP034 AJP069 AJI051-AJI060				
4.6 Customer-supplied product	Non applicable	Non applicable	Non applicable	Non applicable	Non applicable
4.7 Identification, traceability	AJI001-AJI004				

4.8 Process control	AJP09 AJP065 AJP066	AJP013 AJP044-AJP067 AJP022-AJP026	AJP028-AJP029 AJP030 AJP055	AJP016-AJP046 AJP047-AJP048 AJP049-AJP051 AJP052-AJP053 AJP054-AJP063 AJI030-AJI031 AJI054-AJI055 AJI056-AJI057 AJI058-AJI059	AJP070 AJP072
4.9.1 Incoming goods inspection	AJP042-AJP045 AJP029	AJI013-AJI014 AJI046-AJI047	AJI035-AJI036	AJI021-AJI027	AJI046-AJI047
4.9.2 Manufacturing inspection and testing	AJP010				
4.9.3 Final inspection		AJP044 (partly)			AJP070 (partly)
4.10 Control of inspection, measuring and test equipment	AJP07-AJI06 AJP068				
4.11 Inspection status		AJI024-AJI037	AJI042	AJI044	AJI045
4.12 Control of nonconforming products	AJP011-AJP040 AJP043				
4.13 Corrective actions	AJP03-AJP012 AJP041-AJP058	AJI063			AJI063
4.14 Handling, storage, packaging and delivery	AJP056-AJI032 AJI033-AJI034				
4.16 Internal quality audits	AJP002	AJI038			AJI038
4.17 Training	AJP020-AJI015 AJI012				
4.18 Statistical techniques	AJI020				

Reference : AJM001	Version : A	Date : 19.06.92	Page : 57/71

AAF01A

Figure 7.4 Example of procedures attached to a quality manual
(*Source*: Legrand)

Organizational structure

A "technical archiving and reproduction" department centralizes all the technical documents and is in charge of:
- document coding,
- incoming document registration,
- where applicable, normalization of validations and approvals,
- classification (documents in use),
- distribution of documents,
- archiving (documents no longer in use).

Coding and registration of documents

- All technical documents should be coded. The "technical archiving" department keeps a directory of the numbering used (see X 60-200 documentation sheet) and assigns a number to the issuing department.
- On receipt of the document, the person in charge of archiving:
  - verifies the document format, its status (presentation, indexes, dates and signatures),
  - where applicable, affixes the validation and approval signatures with the agreement of the issuer,
  - registers the documents,
  - files the original reference document.

Modification of technical documents

- The modification of a technical document is the responsibility of the issuing department, but it should include the approval of all departments and of all functions concerned with this modification. This is carried out, for example, by means of a modification sheet. Figure 7.5 gives an example of a product modification sheet.
- The person in charge of archiving verifies that the modification has been carried out properly, classifies the modification sheet with the

new original as reference and cancels the old one, which should be preserved in the archives.

Distribution and preservation of technical documents

Document distribution is defined by the issuing department; it is organized by the person in charge of archiving, who must always be aware of how many copies are being sent out and to whom (Distribution tables).

The addressees are expected either not to use or otherwise to destroy outdated copies and to keep "valid documents" (i.e. the last issue (see § 7.1)) which they should subsequently distribute to:
- other internal departments,
- external bodies: suppliers, customers (responsibility of the commercial departments).

7.8 Document storage

The archiving of documents should ensure their preservation under optimal safety conditions and ensure compliance with the traceability rules (see § 6.6).

As a precaution, a copy of each document concerned should be stored (microfilm is often convenient) separately from the originals.

The types of archived documents, their location and the duration of archiving are defined in the procedure in the document control and management procedure (see § 7.2).

TECHNICAL MODIFICATION PROPOSAL N°

REQUESTED BY: Department ———— Name ———— Date ————

EQUIPMENT ☐ Circulator ————
☐ Pump ————

Document number ————
Description of the modification ————

Justification for request : ————

Required application classification : ☐ A ☐ B ☐ C ☐ D

TECHNICAL INFORMATION
Estimated correction time

Expenses ($)

☐ Approval
☐ Refusal

Name ————
Signature ————
Date ————

Estimated design cost

PRODUCTION INSPECTION INFORMATION

Estimated stock ———— Needs

Costs ———— In-Process ————
Finished products ———— Orders in place ————

☐ Approval
☐ Refusal

Name ————
Signature ————
Date ————

1. Application clause	Suppliers orders in progress	Manufactured parts in store	Manufactured equipment in store	855437 ed 3
Class C	Immediate applicat			
Class D	Application for ensuing orders/manufactures			

METHODS INFORMATION
Estimated implementation time
Estimated impact on cost price
Estimated tool costs

Cost price ($)

☐ Approval
☐ Refusal

Name
Signature
Date

PURCHASING INFORMATION
Estimated implementation time
Estimated impact on cost price
Estimated tool costs

($)

☐ Approval
☐ Refusal

Name
Signature
Date

QUALITY ASSURANCE INFORMATION
Estimated tool costs

☐ Approval
☐ Refusal

Name
Signature
Date

FINAL DECISION
☐ Approval
☐ Refusal

Application classification ☐

The requesting party is in charge of this sheet

Distribution to signees

1. Application clause	Suppliers orders in progress	Suppliers orders in progress	Manufactured equipment in store
Class A	←——————Immediate application——————→		
Class B		←——————Immediate application——————→	

Figure 7.5 Example of a product's technical modification sheet

8 Quality Planning

8.1 General

Quality planning is one of the essential components of quality management (see § 1.4). "To manage is to plan" is a familiar expression.

It is also, within this context, one of the bases for quality assurance, since quality confidence depends on the "planned and systematic" actions (see § 1.3) required for "planning what needs to be done" and "doing what has been planned."

Within the context of quality management and for quality planning, these are "the activities that establish the objectives and requirements for quality and for the application of quality system elements" (see ISO 8402 paragraph 3.3). But it is necessary to distinguish between:

- **strategic planning**, i.e. in preparation for the establishment of the quality policy, or in other words, for organizing the quality system by stipulating objectives with the use of a calendar; the corresponding document can be called the "strategic quality plan" (see figure 2.3, note 3).
- **planning for a product, project or service**, which, from the existing system, identifies the specific quality objectives and requirements of a product, project or service.
- **operational planning**, which corresponds to various forms of plans: inspection plans (see § 11.2), surveillance plans, audit plans, etc.

The requirements of the ISO 9001, 2 and 3 standards comprise the preparation "where applicable" of quality plans, but without stipulating their nature and form. They are often the object of specific contractual requirements for important, complex or new products. These plans derive from the quality manual. However, they remain distinct from the manual so that they can be improved separately. A quality plan can never replace a quality manual.

Note that the term "program" has often been used in reference to certain quality documents of this type. ISO 9000 standards use this term only for some specific documents (for example, a program of audits to be carried out as opposed to an audit plan); the word "program" is also used in the general sense of the word.

Contents of the quality manual

It is important that the manual refer to the different existing quality plans, in particular for products or projects, by explaining their role within the series of quality system documents. For example, in a procedure that is referred to in the manual, there is a need to stipulate:
- for what purpose quality plans are established: types of products or services, projects, internal or contractual purposes (quality assurance plans: see § 8.3), etc.;
- how they are drafted, approved and validated;
- the phases concerned: design, manufacture, etc.;
- their content: position of the design reviews, hold points, contributors, etc.

8.2 Strategic quality plans

These types of plans may take varied forms, depending on the policy or quality approach adopted by the organization.
- If the policy aims only to meet contractual or regulatory requirements, or to prepare the organization for certification of its quality system, for example, by meeting ISO 9001, 2 or 3 standard requirements, then no form of strategic quality plan is required: the required

planning, therefore, concerns only the satisfaction of requirements specified for products, projects or contracts (see § 8.3).

However, a strategic plan can be established to define the stages and schedule necessary for corrective actions or quality system improvement objectives.

If the policy is more ambitious, the field is open for various forms of strategic quality plans, for which the quality manual will summarize the principles or refer to them, for example: quality action plans, quality improvement plans, quality indicators establishment plans, quality training plans and quality communication plans.

In this case, the plan will define the guidelines of actions planned and a certain number of objectives with the associated resources. Of course, if this type of plan exists, it will be referred to in one of the sections of the manual: quality policy, quality system, etc.

8.3 Product or process quality plans

Such plans are essential for a project relating to a product or to a series of products or services or new processes (see ISO 9004-1, § 5.3.3). It stipulates, for the different stages ranging from definition of needs through to customer satisfaction:
- quality objectives to be reached,
- material and human resources,
- quality system practices and elements implemented to show that quality has been obtained.

This plan can be established as a complement to the quality manual, specifically for a product or process:
- either because it is required to show the customer that quality requirements, for a particular contract, have been planned and taken into account in an appropriate manner, i.e. thus justifying the customer's confidence; it may then be termed a "quality assurance plan" (see § 2.2.2 and X 50-164);

- or for internal use in the case of a voluntary quality management approach: the quality plan should, as its first goal, contribute to good quality management. It may also be referred to as a "quality management plan" (see figure 2.3).

The term "quality assurance plan" has been chosen because it makes a clear differentiation between the scope of the document and that of the "quality plan," in its general sense, which may contain certain provisions that do not fit into the contractual context and which may contain confidential elements.

8.4 Preparation of a product or process quality plan

8.4.1 Establishment decision

Whether intended for internal use or to meet a customer's demands, a quality plan or quality assurance plan should be established only when the importance of the product, services or process, its newness, complexity or associated safety requirements so justify it. A plan like this can be used for planning:
- design stages,
- manufacturing stages,
- testing or delivery stages.

For contractual purposes, it may be the first document answering an invitation to tender, subsequently modified according to the terms of the order, and changing in keeping with customer requirements as the contract progresses.

In a procedure provided as reference within the quality manual, it is possible to define:
- under what circumstances a quality plan should be established,
- for which products, services or processes it is established, and what format it will take,
- when it is established, and who decides upon it,

- who is responsible for its drafting, examination, approval, updating and archiving.

8.4.2 Drafting, approval and revision

Drafting

In general the quality plan is drafted:
- either by the technical department concerned and responsible for the manufacturing phase together with the quality department, with the approval of the quality manager, or
- by the quality department with the approval of the technical department concerned, or
- for a sub-contracted part, by the sub-contractor, with the approval of the customer's quality department.

The drafting of the quality plan therefore requires teamwork between the quality department and technical departments. This will make it possible:
- with regard to the design, to define the list of the assemblies or sub-assemblies needed to be qualified or formally validated and, to this end, establish the design reviews that need to be planned;
- with regard to production, depending on the planned manufacturing resources, to define:
  . the list of manufacturing and testing processes that need to be qualified,
  . from the successive manufacturing stages through to the acceptance process, the levels of inspections together with the hold points (beyond which the production cannot be pursued without authorization),
- for the overall activities, to stipulate:
  . procedures, work instructions and specifications to be applied,
  . skills and personnel qualifications,
  . reports or various documents to be established in order to demonstrate quality,

- in the case of contractual requirements, the points or matters requiring the customer approval.

When an internal quality plan exists, the quality assurance plan is reduced to an extract of this plan, drawn up by the quality department.

Validation

The quality plan is normally the subject of final validation that consist in checking that all the stages or phases are coherent and suitable to meet product requirements. It is carried out by the quality manager, in accordance with the procedure concerned.

In contractual situations, the quality plans (or quality assurance) can be submitted to an examination and validation by the customer.

Revision

The quality plan should be revised to incorporate changes made to the product, manufacturing process or to quality assurance and organizational provisions. It is a good idea to indicate who is responsible for what. This can be done in easily updated appendices.

8.4.3 Structure and content

The usual starting point is the content of the quality manual and of set of procedures that exists, from which only those provisions specific to the product or service in question are taken, while stipulating, more specifically:

- the successive manufacturing stages and, where applicable, the relationship of the distinct components to each other,

and for each stage:

- the quality objectives, including references to codes, standards and customer requirements,
- the structures established by the supplier:

   . organization chart limited to the plants and the relationships con-
      cerned with the supply;

   . the names of any responsible persons and contacts, especially as
      regards quality;

- the organizational and technical interfaces;
- the procedures applied;
- the testing inspection points or the hold points (with, where applica-
    ble, the intervention of the customer, etc.);
- the records or documents that demonstrate quality;
- the qualifications specific to the equipment, processes and persons;
- for sub-contracted components, the quality assurance requirements
    to be passed on, etc.

The quality plan of a product can be a single document or set of drawings corresponding to different manufacturing stages or to different
components of the product. In the latter case, a component quality plan
can be established by the sub-contractor in agreement with the customer, the "prime contractor" for the overall product.

This document may, depending on the nature of the product or operations and on the particular customer requirements, take the form of:
- either a document with sections that follow the order of one of the
    ISO 9001, 2 or 3 standards; see figure 8.1, or
- one or several documents and diagrams that follow the order of the
    different manufacturing stages or phases, or
- specifically at request of the customer, a "follow-up document" i.e. a
    document completed as operations take place, including the various
    inspection results and the decisions of internal or external partners
    (see § 11.2).

Figure 8.2 gives a simplified example of a product quality plan.

		Interaction with the models for quality assurance		
	Structure of a quality assurance plan	ISO 9001 *	ISO 9002 *	ISO 9002 *
No	Title			
1	**Introductory sections**			
1.1	Summary			
1.2	Scope and field of application			
1.3	Commitment to supplier quality	4.1.1	4.1.1	4.1.1
1.4	Reference and applicable documents			
1.5	Terminology and abbreviations			
1.6	Specific organizational structure	4.1.2	4.1.2	4.1.2
1.7	Management of the quality assurance plan			

2	Quality assurance provisions			
2.1	Contract review	4.3	4.3	
2.2	Design control	4.4		
2.3	Document control	4.5	4.4	4.3
2.4	Purchasing	4.6	4.5	
2.5	Purchaser supplied product and/or services	4.7	4.6	
2.6	Product/service identification	4.8	4.7	4.4
2.7	Product/service traceability	4.8	4.7	
2.8	Process control(1)	4.9	4.8	
2.9	Inspection and testing	4.10	4.9	4.5
2.10	Inspection, measuring and test equipment	4.11	4.10	4.6
2.11	Inspections and test status	4.12	4.11	4.7
2.12	Control of non-conforming product	4.13	4.12	4.8
2.13	Corrective action	4.14	4.13	
2.14	Handling, storage, packaging and delivery	4.15	4.14	4.9
2.15	Quality records	4.16	4.15	4.10
2.16	Internal quality audits	4.17	4.16	
2.17	Training	4.18	4.17	4.11
2.18	Servicing	4.19		
2.19	Statistical techniques	4.20	4.18	4.12
2.20	Other provisions (3)			

(1) This section may refer to the list of applicable production and installation operations
(2) This section may refer to the inspection plan.
(3) If necessary

Figure 8.1 Example of a quality assurance plan structure showing the interaction with the chapters of ISO 9001, 9002 and 9003 standards

* 1987 version

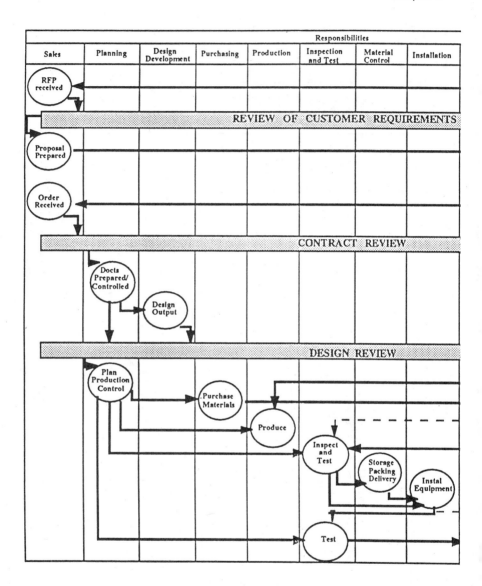

Figure 8.2 Simplified example of a quality plan for a product
(*Source:* ISO/10005)

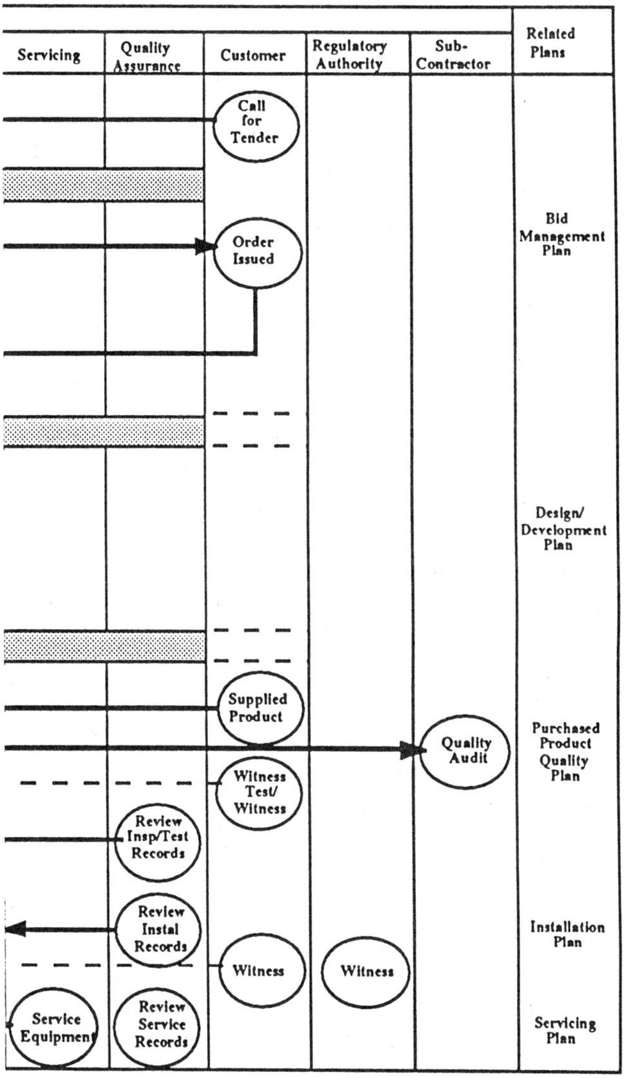
Servicing
Quality Assurance
Customer
Regulatory Authority
Sub-Contractor
Related Plans
Call for Tender
Order Issued
Bid Management Plan
Design/ Development Plan
Supplied Product
Quality Audit
Purchased Product Quality Plan
Witness Test/ Witness
Review Insp/Test Records
Review Instal Records
Installation Plan
Witness
Witness
Service Equipment
Review Service Records
Servicing Plan

9 Quality in Design

9.1 General

Design is a creative activity that starts with the stated needs and existing knowledge, and leads to the definition of a product that meets these needs and that can be industrially workable.

It is one of the quality loop phases, i.e. one of the stages of the project that prepares for the production of a service or product (see figures 6.1 and 6.2). The elements of the quality system relating to design may fall under **project management**.

Defining a product or service in accordance with the stated and implied requirements is the most important stage as far as obtaining quality. This leads to the establishment of documents that can be used to guarantee quality when creating a product or service. Final quality depends directly on the quality of these documents.

It is obvious that better quality is obtained at the lowest cost if, from the start of the design process, the best choices are made and if deviations or nonconformities are detected at the earliest possible stage. Hence the importance of planning the design to fit within the context.of the quality system.

Choices in drafting the quality manual

Obviously, the way the quality manual is drafted with regard to design depends on the nature of the quality objectives of the organization:

- For a manufacturing company with no real design activity that aims to meet the requirements of ISO 9003 or ISO 9002 standards, the manual need not include a "design" chapter.

- For manufacturing activities that start from well-defined product requirements, preoperative production design or a qualification of the final product is necessary. The quality manual may deal with the aspects referred to in this chapter and could also refer to paragraph 10 of the ISO 9004-1 standard.

- For an organization or company dealing in service delivery that aims to certify its own quality system to the ISO 9002 standard, a "design" chapter is not required, but it can be useful to establish the appropriate provisions referring to the content of the ISO 9004-2 standard "Guidelines for services" (§ 6.2 - Design process - and figure 3, even if the external auditor is entitled to refer to it.

- For an organization or company that aims, for example, at certification to the ISO 9001 standard, it is necessary that the manual and specified procedures directly meet the requirements of paragraph 4.4. - Design - of this standard. Paragraph 8 of the ISO 9004-1 standard gives more detailed guidelines.

Some "recommendations on how to obtain and ensure the quality of the design" are provided in the X 50-127 documentation sheet.

Content of the quality manual

In the procedures referred to by the quality manual, the steps taken must be described with regard to:
- design organization:
  - . definition of responsibilities;
  - . interfaces between different technical groups, etc.;
- preparation of design objectives:

    .   planning of design phases (quality plan), which must include the
        design reviews (see § 9.6 - requirement of ISO 9001 standard);
    .   taking into account of standards and regulations;
    .   definition of design quality criteria;
-   design verification:
    .   by persons appointed for their skills,
    .   by appropriate measurements (parallel calculations or testing),
    .   by keeping records of this verification.

9.2 Preparation of design and development

The quality plan

It is necessary to anticipate the establishing of a quality plan (see § 8.3)
at the start of the product design. The plan should be divided into successive phases ranging from product definition through design, development and preparation of the production, tests and delivery (see example of figure 8.2).

The design phases can be divided into stages with pre-defined steps or
hold points and **design reviews** (§ 9.6). These phases generally include
(see X 50-127):
-   a **definition phase**, where the designs translate customer needs into
    technical specifications, the "functional specification," translated
    into "technical needs specification" for contractual use (see NF X
    50-151);
-   a **feasibility study phase**, which aims at showing by means of data
    analysis (see X 50-151 and X 50-152 and documentation sheet X 50-
    153) how it can meet the stated needs by indicating the economically
    feasible technical methods; this leads to an updated functional speci-
    fication;
-   a **draft-projects phase**, which aims at studying the steps that are
    recognized as realizable at the end of the feasibility study phase, with
    the object of proposing one that can be developed;

- a **project development phase**, which is intended to define and qualify the proposed solution and to prepare for product production.

The design phases may lead up to a **production initiation phase** before actual production begins.

For each step, the quality plan defines the participants, the expected partial studies (internal to the organization or subcontracted) that need to be integrated into the overall design phase, the experts to be consulted, the planned verifications, the specifications and procedures applied, the **hold points** and the planned **design review(s)**.

The overall quality plan describes the different stages by clarifying, for each of the responsibilities attributed to the people representing the concerned plants, the links between these plants and the documents established during the design development stage. An extract of this plan - which could be entitled **"quality assurance plan"** (see § 2.2.2) - can be established for contractual use; subsequently, it is generally **validated by the customer** before being applied by the supplier.

Project management

If certain appropriate quality management or quality assurance provisions are to be implemented for projects of a certain size, and referred to in the quality manual, a typical reference could be the "general recommendation for program management specification" RGAéro 00040 (BNAé June 1991) or ISO 10006 "Guide to quality assurance for project management."

9.3 Design input (definition stage)

In an organization, product definition activities include:
- the identification and statement of needs (see X 50-100), stated by the customer and the implied requirements that the organization must define;

- the formulation of this need, which could well be stated in terms of functions and that must lead to the establishing of a "functional specification" (see NF X 50-151) or a document of the same nature; this document will define the assessment and levels criteria of each function.

The definition studies need to focus on aptitude for use and on safety, reliability and maintainability requirements, etc. They must take into account the results of all the marketing activities (or of the contract reviews for contractual situations - see § 6.5).

9.4 Design process

The design development process, taking into account reliability and maintainability requirements, includes:
- design calculations: answers to static and dynamic solicitations,
- choice of materials,
- design of the production resources,
- production process qualifications (see § 11.2).
- design tests and validation.

Figure 9.1 gives an example of a design approach for a pressure vessel.

For a big project, design takes into account all the input data culminating in project realization. Some of these data result from various preliminary studies, and the overall breakdown of these studies should be specified in a document sometimes referred to as the "design flow diagram." Figure 9.2 gives an example of this kind of diagram for an industrial implementation study.

The quality manual should refer to the general provisions adopted to control design development, together with the principal corresponding documents.

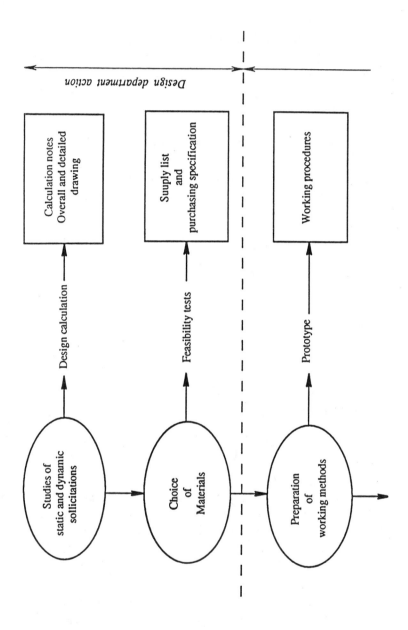

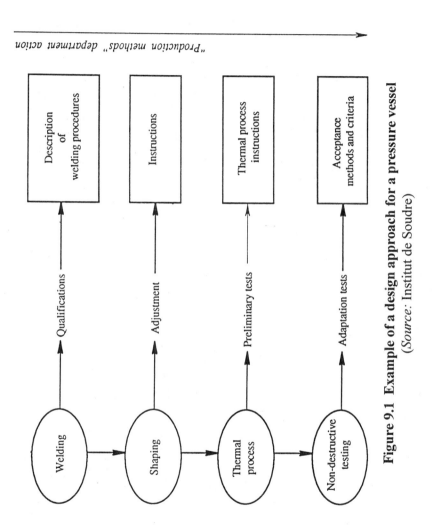

Figure 9.1 Example of a design approach for a pressure vessel
(*Source*: Institut de Soudre)

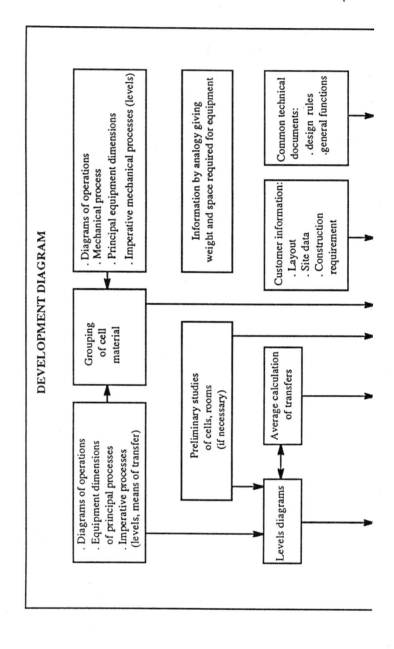

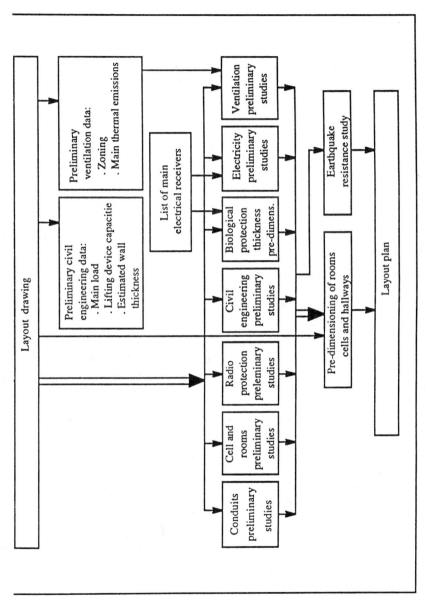

Figure 9.2 Example of a study flow diagram
(*Source:* SGN)

9.5 Design output data

These are documents stating the results obtained at the end of each design and development phase; these results must:
-   meet the requirements relating to the input data of the stage considered;
-   contain or make reference to acceptance criteria;
-   identify the design characteristics that are critical for the correct and safe operation of the product;
-   be validated at the end of the phase.

9.6 Design validation

At the end of each characteristic stage, evidence that the results obtained during each stage are in accordance with the specified requirements must be presented so that the next stage can be tackled confidently. This evidence can be provided:
-   by analytical methods: FMECA (Failure Mode Effect and Criticality Analysis), etc.
-   by qualification tests carried out on samples or prototypes.

The validation process includes assessment of performance, reliability, maintainability and safety obtained under normal operating conditions - i.e., of dependability or RAMS.

9.6 Design review

The design review is a "documented, comprehensive and systematic examination of a design to evaluate its capability to fulfill the requirements for quality, identify problems, if any, and propose the development of solutions" (see ISO 8402, § 3.11).

Such reviews are carried out at the appropriate design phases that have been conceived as part of the quality plan. The participants of each of these reviews must include the representatives of all the functions considered by the design stage, together with any other expert that the customer may deem useful or may demand.

These reviews include primarily:
- integration of basic studies (research and development, patents and choice of licenses, etc.) into realization studies, and their validation;
- where applicable, validation of sub-contracted studies;
- comparison of results with stated needs and various requirements;
- design validation before passing to the next phase;
- analysis of the problems encountered so that they can be avoided in the future (experience feedback).

In contractual situations, and for quality system certification, it is important to keep **up-to-date records of these reviews**, to be able to demonstrate to an auditor that these reviews have been carried out in accordance with requirements, and to show that the resulting decisions have been implemented effectively.

Figure 9.3 gives an example of a design review schedule within a software development cycle.

9.7 Product qualification

The qualification of an entity is the "Process of demonstrating whether an entity is capable of fulfilling specified requirements" (see ISO 8402, § 2.13).

For a product, the process consists in measuring the deviations between what is specified and what has really been obtained. The conformity of the product with the specification results from the following:
- **for design**: the correct translation of the needs expressed into technical specifications;
- for the manufacturing process: a constant aptitude for producing, accurately conforming with the technical specifications (see § 11.2).

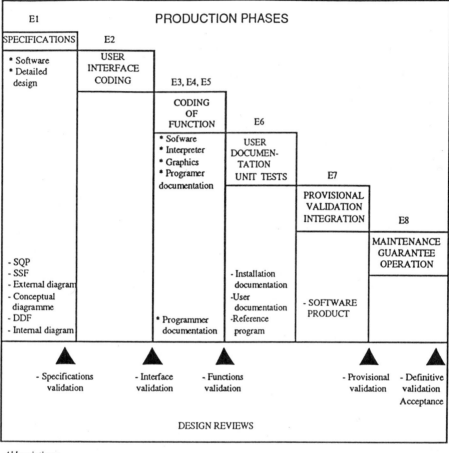

Abbreviations :
        SQP = Software quality plan
        SSF = Software specification file
        DDF = Detailed design file
Conditions for passing from one stage to the next:
   - Only in the case of validation
       E1 ⟶ E2
       E6 ⟶ E7
   - Possible with reservations
       E2 ⟶ E3
       E5 ⟶ E6
   - Not linked to the validation
       E3 ⟶ E4
       E4 ⟶ E5

Figure 9.3 Example of a design review plan for a software development cycle

Product qualification is carried out on innovative products:
- either complete: for a new product,
- or partial: for modification of a fundamental function.

Product qualification must take place prior to initial manufacturing production. It is integrated into the procedure for launching new products.

The qualification process includes:
- within the context of the **development phase** (see § 9.2):
  . the establishing of a technical file,
  . the establishing of a "qualification program," under the responsibility of the technical and quality departments,
  . testing on samples or prototypes (design validation, see § 9.5).
- within a "production start-up" stage:
  . a limited manufacturing run with regular production resources (series zero),
  . inspection and tests in a limited series according to the "manufacturing program,"
  . the "qualification report" established under the responsibility of the technical and quality departments.

Figure 9.4 gives an example of the start-up and qualification process for new products.

9.8 Design changes

The design procedures must stipulate that all the changes and all the design modifications are to be identified in writing and be the object of verifications and approvals by persons appointed for their competence.

For proper control of the modifications, in the case of complex systems and realization tasks, which must start early in the design phase and be

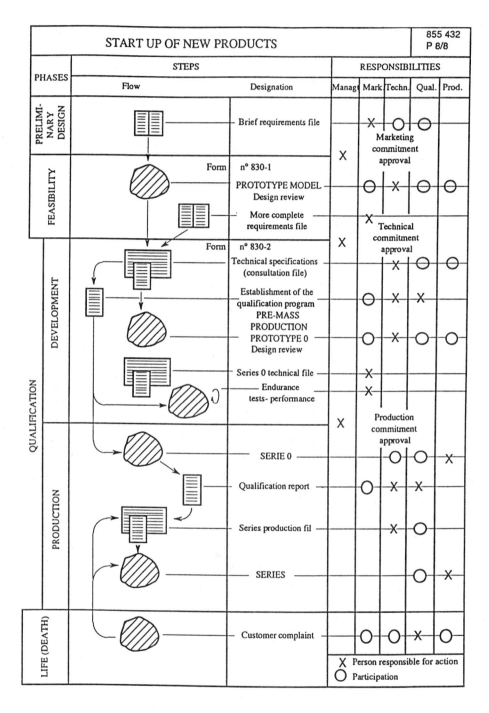

		STEPS		RESPONSIBILITIES				
PHASES		Flow	Designation	Manag	Mark	Techn.	Qual.	Prod.
PRELIMINARY DESIGN			Brief requirements file		X	O	O	
				X	Marketing commitment approval			
FEASIBILITY		Form n° 830-1	PROTOTYPE MODEL Design review		O	X	O	O
			More complete requirements file		X			
				X	Technical commitment approval			
DEVELOPMENT		Form n° 830-2	Technical specifications (consultation file)			X	O	O
			Establishment of the qualification program		O	X	X	
			PRE-MASS PRODUCTION PROTOTYPE 0 Design review		O	X	O	O
			Series 0 technical file		X			
			Endurance tests- performance		X			
				X	Production commitment approval			
PRODUCTION			SERIE 0			O	O	X
			Qualification report		O	X	X	
			Series production fil			X	O	
			SERIES				O	X
LIFE (DEATH)			Customer complaint		O	O	X	O

START UP OF NEW PRODUCTS

855 432
P 8/8

QUALIFICATION

X Person responsible for action
O Participation

**Figure 9.4 Example of an introduction and qualification process
for new products**

pursued throughout the life-cycle of the product - it is advisable to consult the ISO 10007 standard "Guidelines for configuration management."

Some significant modifications can necessitate a partial or complete re-qualification of the product (see § 9.7).

10 Quality in Purchasing

10.1 General

This area concerns all supplies and vendors contracted from outside the organization. It is handled with the cooperation of the various functions of the organization (quality, design, production, purchasing, etc.).

Content of the quality manual

The supply quality is linked to the control of customer-supplier relations, which, within the various functions of the organization, is based on:
- precise and complete expression of needs,
- choice of corresponding requirements,
- evaluation and choice of suppliers,
- placing of orders,
- follow-up of orders, which sometimes necessitates surveillance of suppliers.

The quality manual describes the organizational structure that is set up in this way and the corresponding provisions made (procedures, applicable documents or forms). It stipulates the part played by the quality function in this area.

10.2 Identification of needs and selection of requirements

Quality control starts with the definition of needs and the choice of requirements.

The needs

To establish how, by whom and in which documents:
- the needs are identified; .
- the technical requirements of the product or service delivery are defined, generally translated into specifications and drawings or requirement files (for complex requirements, acceptance levels are included and a functional analysis can be made by referring to the NF X 50-151 standard);
- the quality inspection and quality assurance requirements for quality are chosen;
- the consequent share of responsibilities between customer and supplier is defined in relation to:
  - inspection methods and levels (self-inspection, inspections and tests to be carried out, where applicable, during manufacture, acceptance conditions),
  - records of the results,
  - the supply of a "quality assurance plan" (see § 2.2.2 and paragraph 8.3),
  - the passing along of requirements and responsibilities to the supplier's subcontractors.

The choice of quality assurance

Quality assurance requirements can be by:
- **No contractual clause** (if the non-quality risk is negligible or if product certification to NF EN 45011 standard - by way of the NF stamp, for example - offers sufficient confidence).
- **The choice of a standardized model** for quality assurance (in increasing order of requirements), referring to the ISO 9000 standard:

. **Model 3**: ISO 9003:

"Final inspection only"

To be used when the **final inspection is sufficient** to demonstrate quality (and when product certification is not enough to warrant due confidence);

. **Model 2**: ISO 9002:

"From manufacture to delivery"

To be used when **final inspection is not sufficient** and when the supply does not require a design or development stage*;

. **Model 1**: ISO 9001:

"From design to delivery"

To be used when the supply requires a **design** and development stage* ;

- **The choice of quality assurance requirements specific** to the product, obtained by adjusting the standardized models mentioned above, or conversely, by adding complementary requirements (relating to the software, dependability, etc.).

The recommended approach in determining the choice of model is described and clarified by the decision chart in figure 10.1

The approach adopted, in keeping with this spirit, should be described succinctly in the quality manual or explained in a procedure referred to by the manual.

---

* *Only the 1987 edition of ISO 9001 contains a "servicing" section. In the 1994 revised version, "Servicing" is added to the 9002 standard; ISO 9001 and 9002 become identical except for the quality system requirements for the design of the product or service (see § 11.5).*

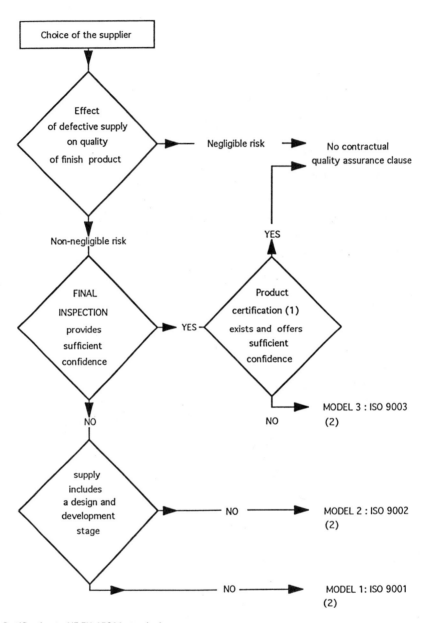

Choice of the supplier

Effect of defective supply on quality of finish product

Negligible risk → No contractual quality assurance clause

Non-negligible risk

FINAL INSPECTION provides sufficient confidence

YES

Product certification (1) exists and offers sufficient confidence

YES

NO

NO   MODEL 3 : ISO 9003 (2)

supply includes a design and development stage

NO → MODEL 2 : ISO 9002 (2)

NO → MODEL 1: ISO 9001 (2)

(1) Certification to NF EN 45011 standard
(2) Possible adjustment: some system elements can be contractually deleted or added
     (Cf. ISO 9000-1, § 8.4.1)

Figure 10.1 Approach for choosing a model for quality assurance

10.3 Assessment and choice of suppliers

The choice of suppliers results from preliminary canvassing and selection based on the evaluation of their ability to meet the quality requirements, as well as deadlines and prices, and to maintain their performance over the course of time. The quality manual has to specify how this choice is made (note that several suppliers can be retained for the same product).

The quality manual also has to answer the following questions:
- who is responsible for preliminary canvassing?
- who is in charge of pre-evaluation and of the ensuing continuous supplier evaluation system, and in accordance with what provisions?
- who decides on the choice of supplier?

For purchasing, the quality manual should indicate the sharing of assignments and responsibilities between the production (responsibility for the product), quality (quality department, quality manager or correspondent) and purchasing departments.

10.3.1 Evaluation

Evaluation generally concerns (see § 13.1) the supplier's ability:
- to meet all the technical, deadline and price requirements, and
- to define and implement an appropriate quality system that will enable the guaranteeing of quality commitments.

The evaluation elements include:
- the supplier's organizational descriptive documents:
  - quality manual, quality plan of the supply considered,
  - preliminary evaluation questionnaire filled in by the supplier (the "standard questionnaire for the supplier's preliminary evaluation" of the X 50-168 documentation sheet can be used or can be a source of inspiration);

- should the occasion arise, knowledge of:
    . quality surveillance results of previous services,
    . the history of the quality policy changes used by the supplier's managing team,
    . evaluations known through other means,
    . approvals or agreements granted by recognized organizations (SIAR, for example),
    . third-party certification of the quality system (e.g., by AFAQ),
    . third-party certification of the product considered (e.g., NF mark),
- results of a complementary audit, which may be decided on if the known evaluation elements are incomplete.

It is advisable to assemble these evaluation elements in an updated report or file.

An example of an evaluation approach followed up by the specifier is provided in figure 10.2.

10.3.2 Choice and approval

The choice of supplier is decided upon with respect to the quality elements resulting from an evaluation as mentioned above. This can be:
- either a "case-by-case choice" made for a given order.
- or the choice of supplier from a list of "accepted," "approved" or "qualified" suppliers (see § 13.5).

Supplier acceptance, agreement or certification are in general established for a determined period of time as specified in the quality manual, in a procedure, or contractually (depending on how specific the supply may be). This approval can be forfeited in the event of unfavorable results or reconsidered when major modifications need to be made to the nature of the products or their specifications.

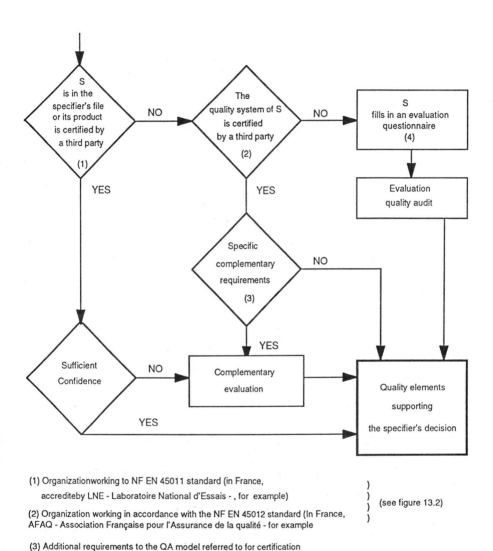

(1) Organizationworking to NF EN 45011 standard (in France,
    accrediteby LNE - Laboratoire National d'Essais - , for example)   )
)   (see figure 13.2)
(2) Organization working in accordance with the NF EN 45012 standard (In France,)
AFAQ - Association Française pour l'Assurance de la qualité - for example   )

(3) Additional requirements to the QA model referred to for certification

(4) For example, the "supplier's evaluation questionnaire" of the X 50-168
documentation sheet can be used

**Figure 10.2 Example of an evaluation approach of a supplier S
followed by a specifier**

Acceptance is, in general, limited to criteria relating to the quality system and to technical quality. Making the choice involves taking other matters into consideration: deadlines, financial issues, etc. (in principle, compliance with costs and deadlines can be included in the overall quality criteria).

10.4 Drawing up of orders

Depending on the size of the organization and its organizational structure (centralized or decentralized), the quality manual, or any procedure referred to in it, must answer the following questions:
- who is in charge of drawing up the order, and on what information criteria (see above: needs, choice of requirements, evaluation elements, whether or not approval or certification is required)?
- how the quality assurance requirements are inserted?
- what procedures are possible for internal order distribution (essentially to verify that the quality requirements meet the needs of the buyer and the organization's internal regulations)?
- to whom are the received orders distributed (internally or externally)?

10.5 Followup, surveillance and receiving inspection

Order followup of supplier surveillance and incoming goods inspection conditions can assume many different forms, depending on the nature of the supplies (repetitive or listed products, or complex products with an element of risk) and on contractual or partnership relationships. The provisions that are made in this area depend on:
- where applicable, confidence in the known quality of the product for its intended use (for example, a product certified to bear a stamp of approval);
- contractual manufacture, followup and incoming inspection conditions, which may vary according to the confidence regarding the supplier's quality system (see ISO 9004-1 standard, paragraph 9.4).

The quality manual describes the provisions applied during the different situations (reference to corresponding procedures should be made).

It should clarify the following points:
- the incoming inspection conditions for listed products - or repetitive products - by suppliers or on delivery;
- the follow-up conditions and possibly the surveillance conditions for complex or sensitive products:
  - . content of required quality assurance plan, where applicable (see § 2.2 and 8.3): specifier's inspector summons points on manufacturing site (hold points), methods of reporting inspections or checks, the documents submitted to prior approval by specifier,
  - . product identification followup (traceability),
  - . deviation correction followup (disposition of nonconformities, production permits, waivers);
- the nature of receiving inspections (100% sampling inspection, see § 11.3.1);
- the settlement of disputes related to quality;
- the receiving inspection records that are necessary as proof of quality (especially the justification of the decisions reached in correcting nonconformities).

The conditions under which the supply and delivery of a finished product need to be followed up or supervised should be pointed out contractually as a function of the way the responsibilities are shared between the customer and the supplier.

The more the customer-supplier relations tend to move towards a true **partnership**, based on mutual confidence, the less the surveillance of the supplier and the delivery inspection are needed. These considerations might well deserve mention in the actual manual.

11 Production (or Process) Quality

11.1 General

This chapter deals more specifically with quality as it applies to the production of material goods. However, in many cases, the delivery of services uses processes that can are comparable in terms of quality control and quality assurance; in the case of contractual requirements relating to quality or to the certification of the quality system of a service delivery organization, one of the ISO 9001 or 9002 standards should be used as reference rather than the ISO 9004-2 standard "Guidelines for Services."

The various sections of these standards are not discussed here; instead, focus is placed on the main principles and documents that the quality manual should describe for a manufacturing activity and that can also often beapplied to a service delivery.

In the following, the term "product" can often be replaced by "service" and the word "production" by "process."

The quality of production depends essentially on:
-   the quality of the documents that define the product: these are documents that result from product qualification at the design stage (see § 9.7) (for a service delivery, similarly, the delivery process can be conceived of and qualified),
-   the preparation of the production operations,

- the quality of the raw materials or basic products,
- the control of the manufacturing processes, starting with their qualification,
- the qualification of persons who produce and inspect the product,
- the maintenance of the equipment and especially the associated software,
- the control of the inspection, measurement and test equipment,
- the identification of the "key points," i.e. the inspection points for a process of particular importance because of the risks entailed regarding the quality of the product or service, the safety of the installations and personnel, and the protection of the environment.

Choice for the drafting of the manual

The way in which the quality manual is written as regards production depends on the nature of the organization's activities and its quality objectives:
- for an organization aiming at production to meet the requirements of one of the ISO 9001 or 9002 standards, it is possible to either:
  . use the same plan as one of these standards for those sections concerning the production, or
  . describe the provisions made in the logical order of the organization's production activities (e.g., incoming raw materials inspection, machining or forming, assembly, testing, etc.).
- for an organization aiming at meeting ISO 9003 standard requirements, the simplest solution is to follow the order of the sections in this standard, which deals only with production activities relative to final inspection, the nonconforming products control and delivery.
- in all other cases, it may be useful to refer to the sections relating to production or to processes control in ISO 9004-1, ISO 9004-2 for services and ISO 9000-3 for software shipping and maintenance.

11.2 Preparation for production

The quality plan

For any activity concerning the production steps ranging from product design to delivery - independently of the quality manual, but by referring to it - a quality plan for the overall development of this activity should be established (see § 9.2) and the production phase(s) of this plan should be considered.

If the activity consists of production alone, a quality plan should be drawn up specifically for this production. This plan will be broken down into successive stages.

For each stage, the quality plan should indicate the parties concerned, the subcontractors who must be integrated into the overall plan, the specifications and procedures applied, the self-inspections and other planned inspections and hold points. It should stipulate the responsibilities allocated to the persons concerned, the points requiring the customer intervention, etc., and the documents established to demonstrate quality.

The "production" quality plan can only aim at a good internal management; it may be required by the customer and in this case may be entitled **"quality assurance plan"** (see § 2.2). This plan contains only the specific quality provisions that refer, for details of the operations, to the "production sheets," or for operation inspection, to the "inspection plans" and, for surveillance, to the "surveillance plans" (for the drafting and approval of this plan, refer to paragraph 8.4.2).

In certain cases - principally at the customer's request - a production quality plan can become a "follow-up document" if it is referred to during the manufacturing development - i.e. completed by inspectors or supervisors who attach their observations and their signature to it after inspection - to collate a final report at the end of manufacture that will contain the results of the main inspections and the decisions reached during the correction of nonconformities (see § 11.3.3). In certain areas, such a document is often called a "list of operations related to manufacture and inspection"; it can be validated by the customer before being applied (see § 8.4.2).

Figure 11.1 is an example of a quality plan produced by a small mechanical parts manufacturer for the nuclear industry and "filled in" by the chosen intervening parties.

Production sheets

Production sheets describe a set of elementary operations that must be carried out to a pre-established introduction rule. This rule is intended to supply the operators with the necessary and sufficient indications to allow them to carry out their work without any mistakes or ambiguities.

A distinction can be made between:
- the **"manufacturing sheet"**: for the product manufacture, it is a list of the successive actions (operations, storage, inspection, transport) necessary for the production. For a service delivery, a "processing sheet" is analogous;
- the **"inspection sheet"**: used for the evaluation of the product process or conformity at different manufacturing stages and indicating the provisions to be made in the case of a defect or nonconformity, so that corrective action may be carried out. A distinction can also be made between:
  - the "process inspection sheet," applied to all the elements needed for product manufacture (equipment, special processes such as welding),
  - the "product inspection sheet," often referred to as "inspection plan."

Figure 11.2 is an example of the manufacturing sheet of a small joints manufacturer. Figure 11.3 is an example of a quality sheet, including the manufacture, process inspection and product inspection.

| SAMC | QUALITY PLAN | | | | | | | | Quality Plan : 1488
Revision : 2
Page : 3/3 | | | | |

MACHINING OF ICE 13000 MW CAPS

BATCH No : LX41
QUANTITY : 100

Key : I = Intervention point
H = Hold point
R = Required statement
NCF = Nonconformity Form
OPERATION No 10 OF THE QUALITY PLAN No 1481

Op. No.	Designation of operation	Specification or procedure	Revision	Notification			Report No.	Date	Signature				Observ.
				QC	CM*	QA			QC	CM*	QA	NCF	
01	Incoming goods inspection and storage	1489 § 3	2	H				30.06.92					
02	Introduction to manufacture	1489 § 4	2	H	H	I		30.06.92					
03	Manufacturing to plan indications	1489 § 5	2	H			9550	08.07.92					
04	Inspections	1489	2	HR			X4001	08.07.92					
4.1	Dimensional inspection	§ 6.1 § 6.2	2				4100						
4.2	Visual inspection	1489 § 6.3	2										
05	Packaging and storage	1489 § 7	2	H				08.07.92					
06	Drafting of end of manufacture certificate	1489 § 8	2	HR				08.07.92					
07	Acceptance	1489 § 8	2	H	H	I		08.07.92					
*	SAMC Contract Manager												

Figure 11.1 Example of manufacturing quality plan
(*Source:* SAMC)

MECALLOY F.D.R.

MANUFACTURING SHEET

N° Date :

N°	Operation	Procedures	Rev.	%	Operator	Date Signature	Time	Delay
	Supply inspection and casting N°	QA/C/1						
	Sampling	QA/C/2						
	Cutting	QA/C/6						
	Drilling Core sampling	QA/C/F						
	Stamping	QA/F/8						
	NC Lath	QA/F/7						
	Milling	QA/F/7						
	Grinding	QA/F/7						
	Visual and dimens. inspection	QA/F/6						
	Dye penetrant	QA/F/3						
	Striping Passivating	QA/F/5						
	Marking	QA/E/1						
	Final inspection							

Customer | Ref. N°

Designation : | Grade :

Station N°	Internal classification	Standard	Issue	Level
Raw material :	In stock	Ordered Qty	Manufact. Qty	Scrap

Source cast number : | Supplier order N° :

DISTRI-BUTION | CUSTOMER | STOCK

SUB-CONTRACT

Heat treatment	Qty	Supplier's Order	Incoming Date	Total
Before machin. □				Time per unit
After machining □				Anticipate time

Manuf.	Stock	Cost Price	Sell Price

PLAN N° | Revision | Establish by : Date: | Modified by : Date:

Figure 11.2 Example of a manufacturing sheet
(*Source:* MECALLOY F.D.R.)

Example of a quality sheet integrating manufacturing sheet, process inspection sheet and product inspection sheet. This example concerns printed circuits.
The various parts used during the printed circuit manufacturing are issued by the incoming goods store, where their quantity and quality is checked.

Manufacturing process	Process inspection	Product inspection
1. Preparation of manufacturing documents		
Documents established by the production inspection department are sent to quality inspection department and the person in charge of the printed circuit wiring workshop.	Verification of documents Possible definition of a particular inspection plan	
These are then distributed to the various production managers: - Component preparation - Implementation - Soldering and retouching - Manufacturing inspection.		
2. Preparation of components		
The person in charge of the " component preparation" defines the lead bending sheets, and drafts the oven drying plan.		
Operation - Oven drying of printed circuits for 12 hours at 100°C bearing in mind that the printed circuits must be soldered within 48 hours of leaving the oven.	Verification by temperature and time sampling.	Verification of the length of the 1st component of each batch on a bare circuit.
- Bending of component leads by batch corresponding to a topological indication.	Periodic verification of the oven. Periodic verification of the jaws and blades of the bending-cutting equipment.	

Figure 11.3 Example of a manufacturing and inspection sheet

Manufacturing process	Process inspection	Product inspection
3. Implementation of the components The person in charge of "implementation" draws up an implementation program, when equipment needs to be inserted semi-automatically or when manual implementation is required. Manual or automatic implementation operation. Any circuits that could not be soldered in the allotted times should be placed in a storage oven (for 3 days at most at 40°C).	Periodic verification of the inserting equipment. Periodic verification of the inserting tools. Verification of the application of the procedures	 The first printed circuit implemented is verified in accordance with the implementation during the mass production, a verification is carried out by drawing out 2/10.
4. Soldering, cleaning, retouching The person in charge sets the "wave soldering" temperatures and speeds. Operation: - The printed circuits are generally "wave soldered" after flux and heating.	Periodic verification - of flux density, - of reheating and bath temperature, - of regulation of speed progress, - of orientation, - of quality of the tin solder.	

Figure 11.3 (continued)

Manufacturing process	Process inspection	Product inspection
- The printed circuits are cleaned in a recycling machine. - The components that could not be "wave soldered" or cleaned are iron soldered. - The soldering touching up tasks are carried out with an iron. - Manual cleaning of the touched up points. - Varnish, screws, etc.	Verification of the quality of the cleaning and flux.	Random sampling check of the soldering quality and calculation of defects as a percentage. **5. Product inspection** Each printed circuit is checked for conformity, soldering and appearance on appropriate testing means. The various defects are retouched and totaled at each work station. After approval of a batch by the quality inspection department, it is sent on to the shipping stores.

Qualification in production

The qualification of a production process consists of formally verifying the ability of a new process to produce continuously a product in accordance with its design definition.

To control production, it is important to start by identifying the processes involved in production: this may have been done during the "development phase" (see § 9.2) and appear in the product qualification file.

Generally, special processes need to be clearly identified, i.e. those that give results that cannot be entirely verified through product inspections and tests and that could cause deficiencies only appearing after the product is used. For example, these could concern welding, heat treatment, non-destructive testing, or activities using software.. These processes require qualification before their implementation.

It is important that the quality manual highlight these different processes and refer to the procedures or documents that permit their control.

Production qualification can include, depending on the activity:
- **equipment qualification** (e.g., welding equipment qualification),
- **personnel qualification** (e.g., welder qualification) - (see § 15.4),
- **technical process qualification** (e.g., argon welding process qualification),
- **incoming product qualification** (e.g., qualification of the filler metal).

Qualification is carried out for a procedure, a specification or a processing sheet, under the responsibility of persons designated in the inspection laboratory, technical department, production department, quality department, etc., or yet again by an external organization (e.g., a licensing authority). It is formalized by updated records that may demonstrate qualification for the external auditor.

11.3 Inspection and tests

The quality manual, in a brief description of this activity, must refer to the written control procedures for inspections and tests and indicate how they are specified in the quality plans specific to the different products. The determination of the scope and nature of the inspections must be by coordination between the quality department, the technical and production departments and, where applicable, in accordance with the customer or sub-contractor: determination of inspection and surveillance levels and corresponding hold points positions or the corresponding notifications.

It may be necessary to identify the need for inspection in terms of statistical inspection techniques; when applying the ISO 9001, 2 or 3 standards, their implementation must be subject to documented procedures.

11.3.1 Incoming goods inspection

The method used to ensure the quality of brought-in materials, elementary and assemblies depends on the importance of the article involved as concerns quality and on the contractual and partnership relations (see § 10.5). This inspection can be carried out by an unit of the organization (often called "incoming goods inspection" or "entry inspection" attached to the production or quality inspection departments:

- on **each article** (100% inspection), or
- on **samples** (standards used: NF X 06-022 and 023) by submitting samples to an "inspection plan" (provided to the supplier so that he knows where he stands); in certain cases (pharmaceutical or agrofood industries), it may be necessary to keep some accepted product samples for a determined period of time, or

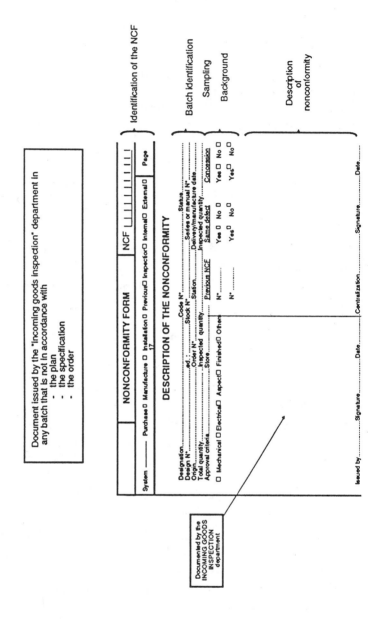

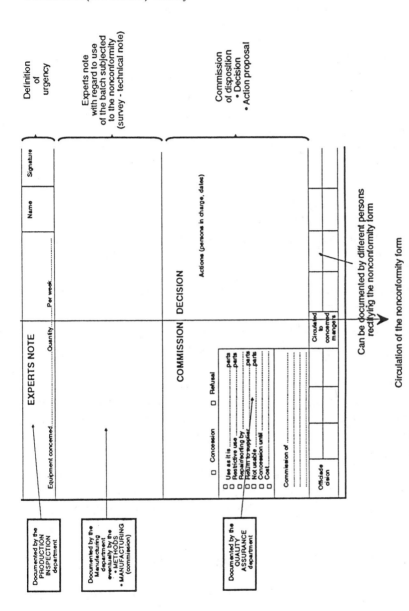

Figure 11.4 Example of nonconformity form used for incoming materials delivery

- on presentation of a **"certificate of conformity,"** established by an independent external organization and recognized or "accredited," called a "third party" (product certification to the EN 45011 standard, see figure 13.2), or
- on presentation of the supplier's **"declaration of conformity"**: the customer deems it sufficient when a supplier has been "accepted," "approved" or "qualified" (see § 10.3.2).

Note: In the last case, the customer may consider that these quality assurance provisions, applied satisfactorily by the supplier, provide sufficient assurance about the supplier's ability to supply a quality product that the customer can forego the burden of incoming materials inspection.

Note that this inspection, which leads to the drawing up of a "report" may be carried out:
- at the source, i.e. at the supplier's company,
- during receiving, i.e. on incoming acceptance at the organization.

Depending on the case, it may be completed by "quality surveillance" at the supplier's company, in accordance with the "surveillance plan" (see § 11.2), and followed by an "inspection report" or a "surveillance report."

In the event of detection of a nonconformity, the incoming goods inspection department establishes a "nonconformity form" which helps the departments concerned to take appropriate action (see § 12.2 and figure 11.4).

11.3.2 Inspection and testing during production

For a given product, it is important to determine in the quality plan the location, the nature of self-inspections and inspections depending on the importance of the product quality requirements, and the ease and cost of production. The inspection data and levels are stipulated in the procedures, specifications or sheets that are available for inspectors and that are referred to in the quality plan.

The information recorded during these inspections can follow the product throughout production and be a part of the final quality evaluation. One of the ways to proceed with this recording consists of "completing" the quality plan, which then becomes a "follow-up document" (see § 11.2).

11.3.3 Final inspection and tests

No product may be delivered before all the actions stipulated in the quality plan have been carried out and the corresponding documentation has been accepted or approved by the designated persons within the organization or by the customer (inspectors, supervisors, quality assurance managers or correspondents, etc.).

The final inspection and test types can be:
- measurement by **individual inspection**, consisting in measuring quality at the end of each manufacturing stage, applied for products with a deficiency level that must be reduced to the minimum.. This measurement is costly and is applicable only when it is non-destructive;
- measurement by **statistical sampling** following an "inspection plan" (see NF X 06- 022 standard). They allow an evaluation of the overall level of quality, but will not easily detect an individual deficiency during the production cycle.

The final inspection must be completed with other actions, in particular with:
- the results analysis of all the inspections carried out during the production, (contained, for example, in a "final report" - sometimes required by the customer),
- more generally the implementation of a quality management policy that optimizes final inspection taking cost/quality aspects into consideration (see chapter 16).

11.4 Inspection, measuring and test equipment

Any organization supplying a product must be able to demonstrate the product's conformity with its stated requirements. It is therefore necessary to establish a **metrological function,** which consists of checking its suitability for use of all the measuring resources used in the organization, together with providing assurance. Its quality manual must, in the appropriate section:

- describe the organizational structure implemented to assure this function with the necessary independence: designation and assignments of the department concerned (quality department, laboratory, etc.);
- refer to the applicable procedures in the different production departments to identify, control and manage the measuring equipment.

Main provisions to be implemented

- Determine the measurements to be made, the accuracy required and the necessary variability and select the appropriate instruments.
- Identify and list all the inspection, measuring and test equipment and resources, including the software used during the installation of the inspections and tests.
- Define the calibration process, its frequency, proper environmental conditions, approval criteria and the measures to be taken in the event of a result that is not satisfying.
- Define the cases where the traceability in relation to standards of reference is necessary.
- Identify each instrument with a mark or tag indicating the calibration status.
- Keep up to date all the calibration records.

A program for the calibration and the verification of the measuring equipment must take into account all the operations to be carried out on the overall measuring equipment.

A distinction is made between internal calibration carried out by persons designated for their competence and the subcontracted calibration by organizations or calibration centers approved by the National Metrology Center.

Some detailed recommendations are to be found in the standards:
- NF X 07-10: "Metrological function in the organization" (completed by the X 07- 11 documentation sheet),
- ISO 10012: "Quality assurance requirements for measuring and test equipment."

11.5   Servicing

The ISO 9004-1 standard (1994) makes a distinction in activities following production (ISO 9004-1, § 16) between the terms "servicing" (§ 16.4) and "after-sales" (§ 16.5). However, the ISO 9001 and 9002 standards (1994) make no distinction concerning after-sales in the section covering post-production activities (§ 16).

In the spirit of these standards, servicing includes services that may be specified internally, or by the customer, to complete the supply of a product, for instance: supply of special equipment intended for product handling or maintenance, instructions for installation on site and integration in its environment, training of product users or services concerning withdrawal of a product. The after-sales section of the ISO 9001-4 standard deals more particularly with failures in use and of corrective action to be assured.

Contents of the quality manual

If the quality manual plan is in accordance with one of these standards, it will be necessary:

- first, in the section "Process control" (9001 or 9002, § 4.9 or 9004-1, § 11), to identify the servicing activities that are or that may be contractually applicable as a complement to the main processes in question;
- in the section "Servicing" (9001 or 9002, § 4.19 or 9004-1, § 16.4), to describe briefly their object and the provisions made (planning, etc.) with reference to the applied procedures, or to justify their absence if necessary.

If the plan of the manual differs from that of the standards, servicing may be dealt with in the logical sequence of activities and processes of the organization (see § 3.3.2, Options 3 or 4 and figure 3.5).

- In the 1994 French revision of the ISO 9000 standards family, the term "prestations associées" (the translation of "servicing") replaces the term "soutien après la vente" used in the 1987 version of the standards.

12 Control of Nonconformities

12.1 General

Nonconformity is the "**non-fulfillment** of specified requirements" (see ISO 8402, § 2.10). However, there are not only specified requirements: within the quality management area, obtaining quality consisits not just of meeting the specified requirements - particularly those of the customer - but also of meeting the implied needs, which, even if they have not been stated, must also be met in the interest of the production manager or the person receiving the benefits, as well as in the interest of the user.

General identification and typology

A distinction can be made in the broad sense of non-fulfillment:
- product or service nonconformities;
- nonconformities or "incidents" during operating activities (installations, transport, etc.);
- procedure nonconformities: non-fulfillment of the quality system requirements;
- anomalies: "deviations in comparison to what is expected" (see NF X 50-120 - the sense can be different in certain areas such as, for instance, in the nuclear industry);
- defects: "non-fulfillment of the intended usage requirements" (NF X 50-120);

- malfunctioning: this word denotes the non-fulfillment of an operation or process;
- failures: "cessation of the ability of the entity to accomplish a required function" (NF X 60-500);
- deviations between the quality aimed at and the quality effectively obtained. This is much less well defined and is often referred to as "bad quality" or "non-quality."

These different types of nonconformities, in a broad sense, are not always appropriate to the organization considered. The manual should make a clear clarification in function of its activity and establish the kinds of nonconformities that are the object of an identification or control provisions.

The provisions to be described

It is important, as much for the good internal management (reduction of the non-quality-related costs) as for quality assurance for the customer, that rigorous provisions are made:

- to identify nonconformities as early as possible during the design, production and servicing process and to take the necessary safeguarding measures;
- to rework the design, or stop production and isolate the nonconforming product;
- to appoint competent persons to:
  . examine the immediate consequences and the decisions needed,
  . research the causes,
  . implement the necessary corrective and preventive actions.

Contents of the quality manual

The manual should describe clearly the organizational measures that are defined in this chapter and should refer to the required procedures, in particular, to the procedures required by the ISO 9001, 2 and 3 standards regarding the **identification** of the nonconforming product or service (detection and marking, if possible, and records), its **isolation** (for a product), **examination, decisions** to be taken, **corrective and preventive actions,** and documentation that needs to be updated (nonconformity sheets or report, customer claims records and their processing, etc.).

12.2 Nonconformities for a product

For a material product, this concerns the control of the nonconforming products. For a service delivery, it applies more to a nonconforming process (see § 12.3).

Origin

For a material product, non-fulfillment can concern a requirement specified:
- in the definition file applied during production;
- in the manufacturing file regarding the manufacturing and inspection conditions;
- in the characteristics defined during the qualification of the product.

It can derive from:
- an identification during the manufacturing and inspection activities;
- quality surveillance (supervision);
- a recording by a quality agent or correspondent;
- a quality audit;
- a customer complaint.

It can include:
- a defective material composition,
- dimensions outside the tolerated standards,
- surface defects (fractures, cracks, etc.),

- internal defects detected through non-destructive examination (X-ray, ultrasound);
- excessively weak characteristics measured during testing.

Classification

After detection, notification of the designated persons, and analysis by the various experts, the nonconformity can be classified, for instance:
- "minor nonconformity": having no consequence on the final operation and use;
- "major nonconformity": likely to have an impact on the operation and use;
- "critical nonconformity": a major nonconformity likely to have an impact on safety and that could incur risk of accident for the users.

Disposition of a nonconformity

This comprises the actions that need to be carried out by the designated persons in order to find a solution. This can be:
- **"correction,"** such as:
    - **"repair,"**: "action taken on a nonconforming product so that it will fulfill the intended usage requirements although it may not conform to the originally specified requirements" (ISO 8402);
    - **"rework"**: "action taken on a nonconforming product so that it will fulfill the specified requirements" (ISO 8402);
- **"regrade"**: i.e. a use for other requirements;
- **"scrap"**;
- **"deviation permit"** or **"concession."**

This applies every time a nonconforming product is detected and it is important that the manual summarize its principle and refer to the procedures, stipulating the methods that apply. These principles are essentially based on:
- the designation of departments or persons capable of reaching the decisions,

- the formalizing of these actions by way of documents: "nonconformity forms," "deviation permit forms," "concession forms," etc., which must be kept in the final quality file (see 11.3.3).

The nonconformity form

Whoever detects a nonconformity (or, at first sight, an "anomaly," (see § 12.1) can bring the right procedure into application by issuing or delegating someone to issue a "nonconformity form" (nonconformity form, see Figure 12.1) in which, after detection, is recorded:
- the identification of the nonconforming product (no. of contract, batch, nature of the product, quantity, material etc.)
- the description of the nonconformity, the circumstances and the acceptance criteria,
- the name of the author or issuer of the form, with date and signature.

By applying the right procedure, the nonconformity form is managed either by the quality department or by the issuing department. However, the quality department verifies the followup of the nonconformity form.

To be able to make a decision that leads to a solution, it is necessary to gather together, from the departments and experts concerned, notes showing:
- the limits attributed to the nonconformity,
- the established origin,
- the anticipated technical repercussions on manufacture and usage operations,
- if the case arises, the financial consequences,
- the recommended decisions to be taken, together with the corresponding preventive and corrective actions.

Appendix A   QA:H1

MECALLOY F.D.R.	QA	QUALITY ASSURANCE MANUAL PROCEDURE	

NONCONFORMITY	N°	Date :	
Customer :	Supplier :		File No
Basic document :			TYPE A: Document B: Organizational structure C: Inspection D: Manufacture E: Other
Description		Acceptance criteria	Product number and quantity

Note and solution proposed by quality assurance department	Quality assurance decision Name - Date - Signature		
	Possible customer decision		
Name - Date - Signature	Name - Date - Signature		
Final Decision :			Name Date Signature
Corrective actions :			
Verification of the final decision performance :	The Quality Assurance Manager		Name Date Signature
Verification of corrective action establishment	The Quality Assurance Manager		Name Date Signature

Figure 12.1 Example of a product nonconformity form
(*Source:* Mecalloy FDR SA)

CERIB	ANOMALY FORM	N° (*)
1 Detection of the anomaly Department concerned : Issued by : Description of anomaly : Released on :		Date : Signature :
2 Disposition of anomalies Person in charge : Description of measures taken : Released on :		Date : Signature :

3 Verification of efficiency of action (if brought back into conformity) Person in charge : Result : positive- negative (1) Released on :	Date : Signature :
3 Follow-up data (where applicable) (Introduction of corrective action, waiver request, internal audit) Released on:	Signature :

(*) Record No. by quality testing coordinator
(1) Delete as appropriate and give details if negative

Addressees: at each stage
- Person in charge of material and test concerned
- Person in charge of quality of testing.

Figure 12.2 Example of an anomaly form for testing
(*Source:* CERIB)

CONCESSION REQUEST

Issuing Department :

Date :
No :
Attribuated by the Quality
Department

Contract   No........

Designation of material......
Quantity.......
Supplier :.....
File No :.....
Order No : ....

Manufacturer :......
Plan No :.......
Allocation No :......

Ref :........
Specification No ......
Delivery No :.........

Object of concession				Name : Signature :

Observations	Department note
	   Accepted
	   Refused
				Name :
				Date :
				Signature :

	Customer note
	   Accepted
	   Refused
				Name :
				Date :
				Signature :

	Quality Department decision
	   Accepted
	   Refused
				Name :
				Date :
				Signature :

Circulation :				

Figure 12.3 Example of a concession request

The final decision is taken in accordance with the quality department and, where applicable, with the customer.

Advice from the various experts, decisions and approvals are recorded on the nonconformity form with the appropriate signatures.

Other types of forms can be used to rectify other kind of "non-satisfaction": anomaly forms, incident forms, deviations forms, etc. (which can, in certain cases of non-satisfaction rectification, precede the declared nonconformity).

Figure 12.2 gives an example of an anomaly form used in a testing center.

The permit or concession request

A distinction is made between two types of permit or concession, before and after production (ISO 8402):
- **"production permit"** or **"deviation permit"**: "written authorization to depart from the originally specified requirements for a product prior to its production,"
- **"waiver"** or **"concession"**: "written authorization to use or release a product that does not conform to the specified requirements."

These authorizations can be given exceptionally for a specific use, for a limited quantity and a limited period of time.

The request in general is stated on a "deviation permit" or "concession" form (example on figure 12.3) where the following is recorded:
- the identification of the product considered,
- the scope and the justification of the request,
- the successive notes of the departments or persons consulted and the acceptance or approval signatures (quality department, customer, etc.).

The request in general is issued by the quality department, which can:

- knowing the problem, answer immediately;
- proceed with certain verifications and give an answer;
- consult a specialized department or external organization before answering.

12.3 Process or procedure nonconformities

These two types of nonconformities concern the non-fulfillment of requirements. They can be considered as similar to the product non-conformities, but they differ in the sense that they are less easily identifiable and measurable than the product nonconformities.

For a process, nonconformity is the non-fulfillment of a requirement concerning this process. This can be, for example:
- a service delivery that does not meet the customer's requirements;
- the development of an operation that deviates from the scheduled operating methods;
- the accomplishment of an administrative approach carried out in a unauthorized manner, etc.

The quality system should be able to define the responsibilities and authority in order to identify these nonconformities, to investigate the causes and to decide which corrective and preventive actions must be taken.

For a procedure, the nonconformity is, for example, the failure to apply a quality system requirement, identified during a quality audit or a quality investigation, relating to the product quality. In this case, it is dealt with as a corrective action following a quality audit (see § 13.3)

12.4 Corrective and preventive actions

It is important to take certain steps to prevent the recurrence of the nonconformity or non-fulfillment by preventing its cause. This can reduce the appearance of other kinds of non-fulfillment. The quality manual should refer to the procedures that are established to this end.

Corrective actions

It is important not to mistake **"correction"** - (which removes the non-conformity but which does not stop it from re-occurring) - with **"corrective action"**: "action taken to eliminate the causes of an existing nonconformity, defect or other **undesirable situation** in order to **prevent recurrence"** (see ISO 8402, paragraph 4.14).

The corrective action implies an analysis of processes and operations that are at the origin of the non-fulfillment, the effective disposition of the nonconformities and of the customer complaints and the analysis of the investigation of the possible causes at each corresponding stage. It can require, for example, improvements to be made with regard to the organizational structure, the procedures and the qualification of material and human resources.

Preventive actions

The preventive action is an "action taken to eliminate the causes of a potential nonconformity, defect or other **undesirable situation** in order to **prevent occurrence"** (see ISO 8402, § 4.13).

It requires the same types of organizational improvement as corrective actions, but it also requires analysis of the potential problems depending on the seriousness of the risks encountered (customer satisfaction, reliability, safety, etc.).

Corrective and preventive actions should be initiated at the start of the disposition of the nonconformity stage or non-fulfillment stage. They should also be followed up by the quality department by means of corresponding documents (or nonconformity forms) to ensure their effective development.

13 Quality Evaluation

13.1 General

The quality evaluation is a "systematic examination of the extent to which an entity is capable of fulfilling specified requirements" (see ISO 8402, § 4.6). The word "entity" can designate here: a product, a service, a process, a person, an organization, a supplier, etc.

This definition applies to cases where the evaluation is carried out in relation to **references** (system of reference); under this definition, these are the specified requirements.

Examples:
- in the case of the organization's quality system evaluation: the references are, for example, those of the ISO 9001, 2 or 3 standards. The result of this evaluation can help to qualify the organization for:
  - . acceptance or approval, for a supplier (see § 10.3.1),
  - . certification or accreditation, in order to recognize its "official" ability (known as "by third party" - see § 13.5).

In these two cases, note that the evaluation of a supplier takes into account the results of the "evaluation audits."
- in the case of the evaluation of a product, process, service or person: the references can be taken from a standard, a specification, or other references.

Word	Quality diagnosis	Quality inspection	Quality surveillance	Quality audit		Quality investigation	Management review
				Internal	External		
What aspect is considered?	Analysis of a company's adaptability to requirements for quality	Compliance with requirements	- Continuous observance of requirements - Supervision of inspections	-Internal quality system to achieve a given goal - A product - A service	Quality system of supplier to meet customer or certifier demands	- An unusual event - An incident - An accident	Suitability of the quality system in relation to quality policy
For what purpose?	Choose suitable quality policy and quality system	Control of process Accepting a product		Verifying the effective application of the quality system to win confidence	Qualifying or certifying a supplier to provide confidence	Casting light on the matter	Improving the quality
Who decides?	The person in charge of the organization	The project leader - Supplier - Customer - Field manager	- Customer - Regulatory authority	The project leader	The customer (or certifier)	The relevant authority or the quality manager (or safety officer)	The top management

	An external consultant (self-diagnosis: a manager)	A qualified inspector (self-inspection: the doer)	An appointed supervisor	An independant audit team: - A qualified lead auditor - Qualified auditor(s) - Experts (according to the needs)	An appointed expert or team of expert	Top management with quality manager
Who acts?	An external consultant (self-diagnosis: a manager)	A qualified inspector (self-inspection: the doer)	An appointed supervisor	An independant audit team: - A qualified lead auditor - Qualified auditor(s) - Experts (according to the needs)	An appointed expert or team of expert	Top management with quality manager
Following what procedure?	By means of a prepared questionnaire	Quality plan or inspection plan	Quality plan or surveillance plan	Audit program and audit plan Implementation of a standard or specified procedure and of a pre-established questionnaire	Review of facts, documents, testimonies, expert appraisals	In-depth assessment of internal audit results and of the quality policy
What presentation of results?	Report	- Report - Conformity attestation - "Informed" quality plan	- Surveillance or supervision report - "Informed" quality plan	Audit report	Investigation report	Report with records of decisions and deadlines
What distribution?	Defined by the person in charge of the organization	According to internal procedures or contract documents	According to internal procedures or contract documents	Top management and relevant managers / The customer (or certifier) and the audited company	Defined by the relevant authority	Decided by the person in charge of the organization
What outcome?	Decision by person in charge of the organization	- Acceptance, correction, waiver, declassification, reject - Corrective and preventive actions	- Acceptance, correction, waiver, declassification, reject - Corrective and preventive actions	Corrective and preventive actions / Acceptance or not of the supplier (certification or not)	Revision, corrective and preventive actions, etc.	- Revision of quality system or quality policy - Updating of the quality manual

Figure 13.1 Terminology relating to quality examination and evaluation comparison table

In other cases, there cannot be a formalized system of references: this is the case for quality diagnosis and certain types of indicators.

Finally, it should be noted that this term covers diverse forms of quality examination ranging from quality diagnosis to management review, via inspection, surveillance or supervision and quality audit.

Figure 13.1 outlines a comparison of varied forms of quality examination and evaluation and the corresponding applications (the audit must not be confused with the investigation, which is an examination but not an evaluation).

Content of the quality manual

Note that the title of the present chapter is not one of the sections that forms part of the family of ISO 9000 standards. However, depending on the scope of the manual and on its structure, certain corresponding provisions might be described, for example, as part of the "quality system," "evaluation of the sub-contractors" and "quality audits" sections.

Most important, especially with respect to an external auditor, is that the quality manual:
- clarify the nature of the implemented quality audits,
- describe clearly the provisions taken for their organization and performance, referring to the corresponding procedures and established audit programs, and also to the way that the auditors are chosen and qualified.

In the quality manual, it is advisable to refer to the indicators and tables used freely for quality management (without any contractual or standardized requirements).

13.2 Quality diagnosis

Quality diagnosis is the "description of the quality status of an organization, one of its areas or one of its activities that is established at its request, and for its own benefit, to identify the strong and weak points, and to propose improvement actions while allowing for the technical, economical and human context" (see X 50-170).

Quality diagnosis is a tool that has shown its importance when applied to an approach to quality system implementation. The diagnosis is in general the first stage of this approach and cannot be improvised: it is often carried out by an expert. It must not be mistaken with the "quality audit" (see figure 13.1).

The diagnosis can also be used at other stages of the quality approach, for example, to extend the quality system to an area to which it does not yet apply, or to take stock of the organization as it stands. It is called a "self-diagnosis" when it is carried out by a manager of the organization concerned.

For more details, consult the X 50-170 documentation sheet. Standardized provision on this subject is not available within the family of ISO 9000 standards.

13.3 Quality audits

The audit is a "systematic and independent examination to determine whether quality activities and related results comply with planned arrangements and whether these arrangements are implemented effectively and are suitable to achieve objectives" (see ISO 8402, § 4.9).

For the quality system, this is an essential quality assurance activity intended to demonstrate the adequate implementation of this system (see Point 7 of figure 1.2). The quality audit - one of its purposes is to evaluate the improvement need or the corrective action - can take various forms, as can be seen from the following:

Word :	Certification	Accreditation	Qualification	Approval
What does it concern?	- Product - Service - Process - Person - Organization	- Certification body - Laboratory	- Product - Service - Process - Person - Organization	- Product - Service - Process
What for?	Written certification of conformity to specified requirements	Formal recognition of the ability of: - a certification body - a laboratory, for a given type of test	Demonstration of the ability to meet the requirements, or of the competence to carry out a task	Authorization to market or to use in an objective way or in defined conditions
Who delivers?	A certification body "third party" independent from the parties in question	A recognized accreditation body	- A responsible internal person or a person external to the organization - A professional organization capable of qualification	A competent and recognized authority

Following what procedure?	- Normative references (1) - Organization: application form, evaluation audit - Product: tests - Person: examination	- Normative references (1) - Application form - Evaluation audit	From the documents of reference: - Standards, specifications, - Procedures, - Files	Official test or verification procedure
Issued document	- Certificate - List of accessible certificates - For the products: mark or stamp	Accreditation Accessible list	Technical documents (aptitude report, test report) - Attestation	Official document
Period of validity	- Determined by the regulation of the certification body - Renewable and subject to suspension or withdrawal	- Determined by the accreditation body - Renewable and subject to suspension or withdrawal	Determined by the procedure	Determined by the procedure

(1) In Europe, the series of EN 45000 standards (in France, NF EN 45000 standards).
(2) In France, AFAQ, for example, for the certification of quality systems of an organization.
(3) In France, COFRAC (the French accreditation committee)

Figure 13.2 Comparison of the different words used for ability recognition

Quality system audits

- **Internal quality system audits**

These audits are important:

- . for a free internal approach to quality assurance:
     see section 5.4 of the ISO 9004-1 standard;
- . for the supplier, to demonstrate to a customer that the quality
     system meets customer requirements (the requirements of one
     of the ISO 9001, 2 or 3 standards, or specific customer re-
     quirements);
- . to demonstrate to a certification body that the quality system
     fulfills these same standards.

It can be seen from the new version of the ISO 9001 and 9002 standards (§ 4.7) that the **establishing and updating of the written procedure are required for the planning and the implementation of
internal quality audits**. In the new 9003 standard, the requirement
concerns only the programming and the introduction of quality audits.

- **External quality system audits**

These audits play a role in supplier evaluation but they are only one of
the resources used for this evaluation (see § 10.3.1) and the evaluation
audits needed depend on the type of product and on the type and extent of the control carried out by the organization on its sub-contractor
(nature of the customer-supplier relationship)

Consequently:

- . for a free approach to quality management, the role played by
     these audits depends on the supplier acceptance policy and on
     the available auditor resources.
- . when applying the ISO 9001 and 9002 standards, no direct re-
     quirement is available for external audits; however, there are
     overall quality requirements applied by the suppliers to their
     sub-contractors (see § 4.6.2 of these standards).

It can be seen from the new version of the ISO 9001 and 9002 standards that reference is made to the ISO 10011 standard, which gives guidelines for the implementation of quality systems audits.

Quality audits of the product, service or process

These are audits that can take various forms depending on the activities concerned. Methodical and independent examination essentially does not focus on the quality system at this stage, but instead it focuses more on the requirements of a technical system of references corresponding to the quality loop activities such as design, production, etc., or, more particularly, to quality plan phases.

These audits can also consist of internal audits, carried out within technical or manufacturing departments, or they can be external audits carried out at the sub-contractor's. For the sake of efficiency, they should be carried out using a procedure similar to that of the auditing system but suited to their more technical characteristics.

13.3.1 Audit management

The ISO 9001, 2 and 3 standards - new version - require that the internal audits be "scheduled on the basis of the status and importance of the activity to be audited " and "carried out by personnel independent of those having direct responsibility for the activity being audited." They refer to information provided on the subject by the ISO 10011 standard.

Choice and qualification of auditors

As stated in the ISO 10011 standard - 3 (X 50-136-3), it is recommended that an "evaluation committee" be established that will qualify the auditors as per ISO 10011 - 2, after appropriate selection and training. The desirability of implementing this recommendation depends on the size of the company and on the organization of its quality system (see § 5.2.2). The following situations might be considered:

- in an organization that already has a quality committee, this committee will be able to play the role of the "evaluation committee."
- when an organization has quality "coordinators" or "correspondents" at its disposition, these persons, who in general carry out their functions on a part-time basis, will be able to be selected to form "a reserve" of qualified auditors. The same applies for certain persons working within the quality department or management.
- for small organizations, it may be necessary to train one or two auditors chosen at a level close to the organization management. It is possible that the head of thie organization will decide to carry out the quality audit him- or herself; however, in this case, the person will have to prove a willingness to follow through on the conclusions of this audit during a "management review" (see § 5.4) that will bring together all his close staff members (or persons delegated to carry out this assignment).

With regard to the auditors selection criteria, a distinction must be made between internal and external audits:
- for an internal audit, quality experience is not the main requirement: it is the level of ability to carry out the analysis and the synthesis that is predominant,
- for an external audit, there are many possible situations; a choice must be made between:
  . a system audit, done by a certified quality auditor (a quality assurance manager, for example);
  . a product or project audit, done by the project manager or one of his or her assistants.

Audits program

The Quality Director or Manager must establish an audit program (generally, annually) to cover all the areas of the organization concerned for a given period of time (generally one or two years). **An external auditor will be able to verify that this program exists and is followed.**

13.3.2 Audit performance

Audits are generally carried out by an "audit team" made of a "lead auditor," (designated, for example, by the Quality Director or Manager) or of one or several auditors, including, where applicable, an expert in the audited activity.

Obligation of independence

To audit an area of the organization, the auditors must be chosen from a different department or sector. For example, when an organization has of a network of quality "correspondents" in different areas, each of them, if qualified, may be chosen to audit an area outside of his or her own.

Preparation, development and report

To be efficient, the audit must be carried out by applying a procedure that must specify:
- any previous agreement with the audited party: date, time, fields of application, system of references;
- preparation: establishing of a "audit plan" and of a list of questions;
- a formalized development: opening meeting, examination of documents and field examination, conclusion meeting;
- a formalized report containing observations on nonconformities and remarks made on points to be improved;
- a followup of required corrective actions.

For more details, especially if a uniform layout for the front page of the audit report is needed, see the ISO 10011-1 standard, together with the X 50-165 documentation sheet.

13.4 Quality indicators and quality performance charts

A "quality indicator" is "chosen information, associated with one of the phenomena, intended to observe periodically development with regard to 'quality objectives.'" A "quality performance chart" is a "synthetic visualization that characterizes the situation and the development of quality indicators" (see X 50-171 documentation sheet). This documentation sheet provides a method of structuring thoughts and facilitates the implementation of tools.

They are tools that play a part in improving quality within the area of a free-willed management approach to quality (see § 16.2).

A distinction can be made, for example, between:
- indicators of nonconformities related to products and to the quality system itself,
- indicators of malfunctions related to processes that are essential for the quality of the product,
- indicators of internal fulfillment (reduction of non-quality-related costs, personnel satisfaction) and customer satisfaction.

It is advisable to refer in the quality manual to the indicators and quality performance charts that have been established for such an approach.

13.5 The forms of evaluation recognition

The various evaluation forms of an "entity" (product, service, process, person, organization or company, etc.) can make it possible to qualify this entity formally by establishing a document recognizing its ability. For certification, accreditation and approval, the certificate is issued by an independent organization referred to as "third party." The word "qualification" has a more general meaning (see § 9.7 for the products, and § 11.2 for persons and processes, etc.).

Figure 13.2 compares the ways of using the different words relating to the recognition of ability.

The certification bodies and accreditation bodies are assessed and accredited by applying a standard (in Europe, the EN 45000 family of standards). Figure 13.3 gives an example of standard requirements for the contents of the quality manual for certification bodies performing quality system certification.

Figure 13.3 Example of required contents of a quality manual for accreditation of a certification body operating quality systems certification
(*Source:* Draft revision of EN 45012: 1989, § 4.4.3)

The quality system should be documented in a quality manual and associated quality procedures, and the manual should contain or refer to at least the following:

a)   a quality policy statement;

b)   a brief description of the legal status of the certification/registration body, including the names of its owners, and, if different, the names of the persons who control it;

c)   the names, qualifications, experience and terms of reference of the senior executive and other certification/registration personnel, affecting the quality of the certification/registration function;

d)   an organization chart showing lines of authority, responsibility and allocation of functions stemming from the senior executive and in particular the relationship between those responsible for the assessment and those taking decisions regarding certification/registration;

e)   a description of the organization of the certification/registration body, including details of the single committee, group or person identified in clause 4.2 c), its constitution, terms of reference and rules of procedure;

f)   the policy and procedures for conducting management reviews;

g)   administrative procedures including document control;

h)   the operational and functional duties and services pertaining to quality, so that the extent and limits of each person's responsibility are known to all concerned;

i)   the procedure for the recruiting and training of certification/registration body personnel (including auditors) and monitoring their performance;

j)   a list of its subcontractors and details of the procedures for assessing and monitoring their competence;

k)   its procedures for handling nonconformances and for assuring the effectiveness of any corrective actions taken;

l)   the policy and procedures for implementing the certification/registration process, including:

i)   the conditions for issue, retention, and withdrawal of certification/registration documents;

ii)   controls over the use and application of documents employed in the certification/registration of quality systems;

iii) the procedures for assessing and certifying/registering suppliers' quality systems;

iv)   the procedures for surveillance of certified/registered suppliers.

m)   the policy and procedure for dealing with appeals, complaints and disputes;

n)   procedures for conducting internal audits based on the provisions of ISO 10011-1.

14 Quality Costs

14.1 General

The principal concern of any organization must be to supply products or services, or to carry out activities, that meet the expressed and implied needs at the lowest cost. In France, for example, the quality-related costs of an organization - which include non-quality-related costs - vary between 5 and 25% of the turnover, depending on the organization and on the implemented evaluation systems. The Ministry of Industry has said: "400 billion francs [~$70 billion] are put into play each year and are lost in retouching, wastes and rejects."

Cost assessment of the overall malfunctions, anomalies, nonconformities, i.e. non-quality, is a fundamental aspect of quality management because, through actions implemented for quality improvement, it leads to:
- controlling and reducing quality-related costs,
- subsequently developing or maintaining production flows.

The benefit of such a policy is to **restore and increase profit margins**.

Does this subject need to be included in the quality manual?

- If it is a "quality assurance manual," i.e. a document drafted for contractual use - or for the certification of the quality system - by

applying the ISO 9001, 2, or 3, there is no requirement: Note that
cost-related subjects are not mentioned in these standards.
On the other hand, if it is a "quality manual" drafted for an effective quality management approach - i.e. primarily a document for
internal use, a tool designed to help the organization members improve quality, assess non-quality-related costs and reduce them -
then the subject indeed merits inclusion in the manual.

However, note that some customers - especially in the case of a supplier–customer partnership - are not insensitive to the way in which
the non-quality-related costs are controlled by the suppliers, and a
quality manual given to a customer may well describe certain provisions to be made. However, when this is done, confidential subjects
should be excluded.

In this chapter, focus will be put on definitions and principles extracted from the existing standards and current standardized developments as a way of evaluating non-quality-related costs. This is the fist
step towards a reduction of these costs and the improvement of the
"economic factor."

14.2 Definitions related to quality and to non-quality costs

At present, no overall harmonized definition is available in this area;
however, the following are worth considering:

"Non-quality" is "the overall deviation between the quality that is
sought and the quality that is effectively obtained." This difference
can be more or less completely evaluated in economical terms (see
Figure 14.1, NF X 50-120). Non-quality is not defined in the ISO
8402 standard.

"Quality-related costs" are "those costs incurred in ensuring and
assuring satisfactory quality as well as the losses incurred when satisfactory quality is not achieved." Some losses might be difficult to
quantify but can be considerable, such as loss of customers (see ISO
8402).

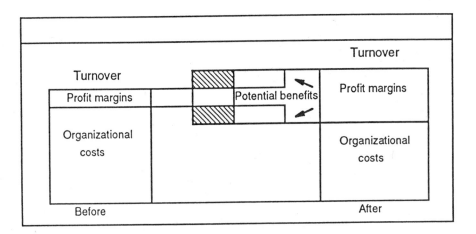

Figure 14.1   Benefit of a quality policy
(*Source:* X 50 126 documentation sheet)

The **"quality losses"** are "losses caused by not realizing the potential of resources in processes and activities." Some examples of quality losses are the loss of customer satisfaction, loss of opportunity to increase the value-added factor for the benefit of the customer, the organization or society, as well as waste of resources and materials (see ISO 8402).

The **"cost of conformity"** is "the cost to fulfill all of the stated and implied needs of the customers in the absence of failure of the existing process." (see ISO/CD 10014 and ISO 9004-1).

The **"cost of nonconformity"** is the "cost incurred due to failure of the existing process" (see ISO/CD 10014 and ISO 9004-1).

The **"economic factor"** is the relation "between the gain and the costs used as a measuring basis to maximize the possibilities of profit for the organization" (see ISO/CD 10014).

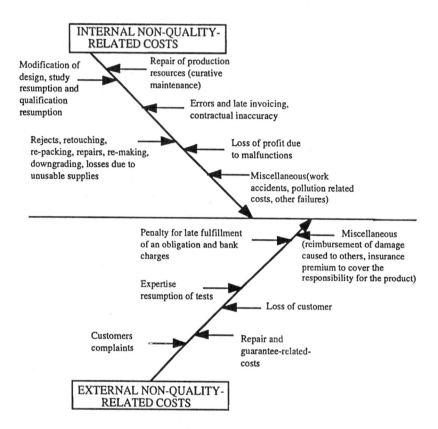

Figure 14.2 Determination of quality-related costs

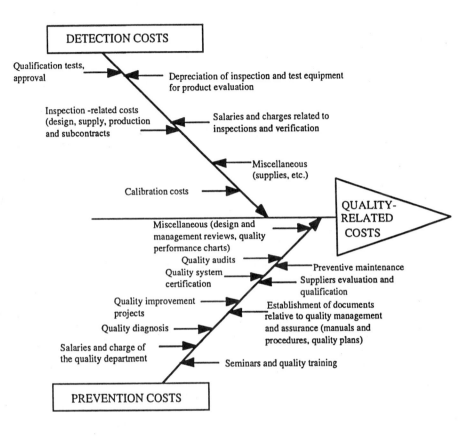

	Monthly cost			Yearly cumulating		
	Objective	Obtained	% of sales	Objective	Obtained	% of sales
1. *Detection costs*						
1.1 Salaries and charges related to Products quality inspection						
1.2 Calibration costs						
1.3 Depreciation of inspection equipment						
1.4 Testing and qualification costs						
2. *Prevention costs*						
2.1 Salaries and charges of quality management						
2.2 Seminars and quality training costs						
2.3 Suppliers evaluation and qualification						
2.4 Internal audits						

3. *Internal non-quality costs*	
3.1 Design modification	
3.2 Repairs	
3.3 Rejects	
3.4 Miscellaneous	
4. *External non-quality costs*	
4. Penalties for late fulfillment of an obligation	
4.2 Return costs	
4.3 Expertise	
4.4 Miscellaneous	
OVERALL TOTAL	

Figure 14.3   Example of a monthly report of quality-related cost evaluation

The "contribution margin lost" concerns the "specific and measurable degradation of turnover or cost which can always be imputed to work that is badly conceived or executed" (see X 50-180-1).

14.3 Determining quality costs

The quality-related costs (sometimes called non-quality-related costs) can be defined as follows:

Internal non-quality-related costs: costs incurred when a product does not meet the quality requirements before leaving the organization.

External non-quality-related costs: costs incurred when a product does not meet the quality requirements after delivery.

Detection and inspection costs: costs incurred to verify the conformity of the product with the quality requirements, i.e. to finance non-quality-related research work.

Prevention costs: human and material investments made to reduce the non-quality-related risks: these include the costs of quality system implementation and maintenance and particularly of quality assurance activities.

Figure 14.2 shows a diagram for an approach of determination of quality-related costs.

Figure 14.3 gives an example of a monthly report of the evaluation of these costs.

For small and medium-size enterprises, it is considered that these different costs can be defined as follow:
- detection + prevention: an average of 10 to 20%,
- internal and external non-quality: an average of 80 to 90%.

A quality approach must stress prevention, by accepting that an increase of prevention-related costs will be an investment that will quickly be covered by the reduction of internal non-quality-related costs.

14.4 Improvement of economic factor

The economic factor is expressed as:
- for a profit making organization:

$$\frac{\text{Benefits in terms of profit}}{\text{Costs}}$$

- for a non-profit making organization:

$$\frac{\text{Benefits in terms of the perceived service value}}{\text{Costs}}$$

To improve this factor, it is necessary to adopt a structured approach. For example, the potential improvements can be outlined in a tree-chart: see Figure 14.4.

For certain details, see the following standardized documents:
- NF EN ISO 9004-1 standard - Paragraph 6.
- X 50-126 - "Evaluation guidelines of non-quality related costs."

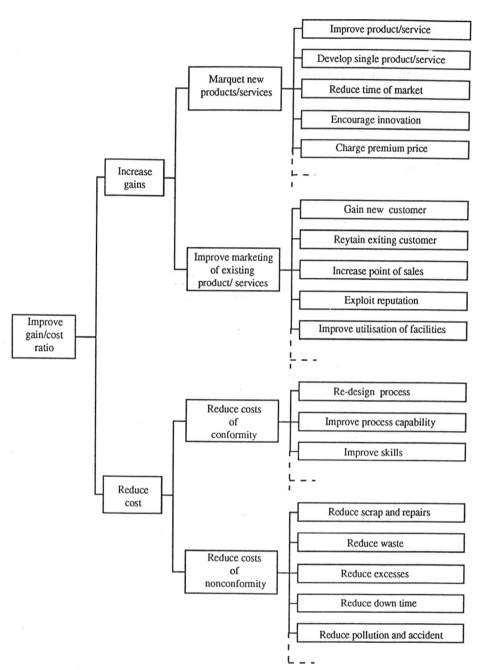

Figure 14.4 Example of an approach for improvement of gain/costs ratio
(*Source*: ISO/DIS 10014)

15 Training

15.1 General

The **competence of the personnel** is a fundamental element of all quality system; this is the reason why training is:
- an important requirement for quality assurance (see section 4.18 of the ISO 9001, 2 and 3),
- an essential recommendation for an effective approach to quality management (see chapter 18 of the ISO 9004-1 standard).

In all cases it is one of the basic starting points for a progressive approach and is therefore one of management's responsibilities, as part of its quality policy and objectives, to anticipate the necessary human resources. This approach must be put into effect by the joint efforts of the "human resources" (see § 5.3) and "quality" functions (see § 5.2.2).

Content of the quality manual

The quality manual should describe the organization adopted for personnel awareness, training and qualification by referring to the procedures established in this area. This includes principally the **identification of the training needs** corresponding to the activities affecting quality, together with the specialized tasks that require a particular qualification (i.e. a process demonstrating a capability).

The ISO 9003 standard does not require procedures.

15.2 Training needs

In order to identify the training needs - initiation or awareness, specialization, re-deployment - it is necessary to give attention to all the categories of personnel, i.e.:

- at a hierarchical level, to the management personnel as well as the production personnel, principally with regard to the different aspects of the quality system applied;
- to the personnel that have been newly assigned or assigned to new activities;
- to temporary personnel;
- to the technical functions, by considering the different activities of the quality loop (see § 6.1): design, production, inspection, maintenance, after-sales servicing, etc.;
- to the quality, safety, environment, supply, computer functions, etc.

It is recommended that these needs be outlined in a **training plan** submitted to general management approval. In a situation of contractual requirement or of certification of the quality system, this plan can be presented to the auditor.

15.3 Awareness and training for quality

Quality awareness and motivation

This is the first approach taken in order for each member of the organization to understand the importance of well-prepared and well-performed work and the consequences that result in the satisfaction of the customer, cost control and a prosperous organization. Within a policy referred to as "total quality management" (see 1.4), its purpose is to induce awareness and involve all of the personnel, from the highest to the lowest corporate levels.

It is recommended that a **quality awareness program** be established, for example, covering a period of 12 to 18 months, that will address all the personnel throughout all levels of the organization. This awareness encouragement is generally carried out on-site by a person responsible for the quality function or by a training organization.

Quality training

This is the training of all those who carry out a "quality function" (see § 5.2.2 and 6.2). This training includes a share of theoretical knowledge and a share of field experience:

-   The theoretical knowledge (quality management and assurance, inspection methods, statistical techniques, quality tools, etc.) can result from technical or university studies or from participation i specialized seminars.
-   Field experience is essential for the quality function to be efficient. It is used for implementing a quality system or for carrying out a quality auditor's function and should be apprropriate to the function carried out, for example:
    .   for a production inspection function, it is desirable to have some production experience;
    .   for a quality correspondent function (often part-time) in a design department, a design engineer is often chosen;
    .   for a quality manager function, it is highly desirable that the person chosen will already have held a management position in a factory.
    .   for a quality auditor function, to have a diversified industrial experience alternating the design, production and management function is desirable; this function must be carried out in parallel with other responsibilities related to operating activities.

The best quality training is obtained by a career management that takes into account both theoretical training and field experience. With this in mind, it is recommended that the "human resources" function manage quality training and that a record of the successive training be kept in the personnel files.

Name:

Department: Function:

Evaluation elements by unit value (UV)

Qualification criteria	Needed requirements	UV number	Observations	Attributed UV
1. General instructionlevel :	A- levels: A levels + 2 or 3 years A levels + 4 years or more: One year studies in a subject relating to quality:	2 3 4	Minimum required: 2 UV Maximum: 5 UV	
2. Audit training	4 days training, by a recognized organization:	2	Minimum required: 2 UV	
3 Industrial experience • Professional: • Quality:	4 years minimum: Technical responsibilities: Quality or quality assurance function responsibilities:	1 + 1 per year 1 per year	Minimum required: 2 UV Maximum: 3 UV Minimum required: 1 UV Maximum: 2 UV	

		1 per audit	Minimum required: 4 UV Maximum: 6 UV	
4. Quality audit experience:	Experience of 4 audits carried out in 20 days minimum			
5. Personal abilities: • Personal qualities: • Experience, seniority:	Appreciated by the person responsible for the audit and other crediting entities		Minimum required: 2UV Maximum: 4 UV	
6. Ability to carry out audits in the following languages:	English: Other languages:	1 1	Minimum required: 1 UV Maximum: 2 UV	
	TOTAL		Minimum required: 16UV Maximum attributable: 24 UV	

Attestation of the qualification

	Dates	Evaluation committee representative		Expiration date
		Name	Signature	
1st qualification:				
Successive qualifications:				

Figure 15.1 Example of a qualification form for a quality auditor
(Completed by an Evaluation committee)

15.4 Qualification of personnel

Some specialized activities require special qualification of the personnel after a specific training (see § 13.4): this is a formal recognition of ability following a process that can be required by a standard or regulation (see § 13.5 and figure 13.2):

- **"qualification"** designates the demonstration of ability in applying an organization's internal procedure or a standard, for example, for a welder or for a quality auditor;
- **"certification"** designates the recognition by an independent organization (third party) of conformance to a specified requirement, for example, the "COFREND certification" for a non-destructive testing representative (X-ray, ultrasound, etc.).

Certain activities such as the operation of cranes or other heavy equipment or work on electrical equipment require "accreditation." This word designates a formalized authorization by management or a recognized licensing organization.

It is necessary, in these documents or procedures referred to in the quality manual, to define the **qualification, accreditation** or **certification** needs of specialized personnel and to clarify the methods needed to obtain these recognized abilities.

Figure 15.1 gives an example of a quality auditor qualification form, based on some of the evaluation criteria recommended in the ISO 10011-2 standard.

16 Quality Improvement

16.1 General

Quality improvement is one of the components of quality management (see § 1.4 and figure 1.1). It includes, in the terms of the ISO 8402 standard, the "actions taken throughout organization, to **increase the effectiveness and efficiency of activities and processes** to provide added benefits to both the organization and its customers."

Several types of actions contribute to quality improvement:
- **quality assurance** related to quality control improve the processes (quality audits + corrective actions + preventive actions = improvement);
- the **"hunt" for malfunctions**, which can be translated by the reduction of non-quality related costs;
- **prospective actions**, for planning internal and external needs;
- the **choice of better tools and techniques** and the training needed for their use.

The first aspect, "quality control + quality assurance," can be the object of requirements, and a quality auditor can verify whether they have been met: this is why these requirements are at the basis of, principally, the ISO 9001, 2 and 3 standards. However, these standards do not include a "quality improvement" section.

On the other hand, other aspects that are not as easy to audit nonetheless have a great importance within an effective quality management approach. The quality improvement is principally the object of the new section 5.6 of the revised 9004-1 standard (which refers to ISO 9004-4 for further details).

Should this subject be treated in a quality manual?

In the same way as for quality-related costs (see § 14.1), there are no requirements when applying the ISO 9001, 2 and 3 series.

For an effective quality management approach, it is advisable to describe the improvement principles and methods applied and to refer to the principal improvement tools to be used.

16.2 General principles for quality improvement

It is the responsibility of upper management to create an environment favorable for quality improvement and management must orient the quality improvement effort by defining certain objectives and **action plans that will involve all the personnel.**

Examples of environmental factors

- Leadership by all the hierarchy;
- Concentration on internal and external customer satisfaction;
- Involvement in quality improvement by the entire customer/ supplier/subcontractor chain;
- Promotion of teamwork coupled with respect of individual achievement and acknowledgment of merit;
- General use of quality indicators and quality performance charts (see § 13.4 and X 50-171 documentation sheet);

Example of objectives

- to make personnel aware of the quality principles and to train them progressively (see § 15.3);
- to identify the activities for which quality indicators can be implemented and to plan for this implementation;

- to implement quality-related cost measurement and to reduce non-quality-related costs (see § 14.3 and figures 14.2 and 14.4);
- to establish quality improvement plans in the different sectors;

The corresponding actions have to be related, especially to the quality of products, services or processes affecting safety, reliability, maintenance, environment and more generally to the customers' satisfaction at all levels;

16.3 Quality process improvement

In the ISO 8402 standard, a process is a "set of interrelated resources and activities which transform inputs into outputs." Within an organization, it can be considered that the different technical and administrative activities form a group of processes, and that these processes are a succession of tasks carried out with human and material resources, information and by appropriate methods or procedures. Each process is characterized by measurable inputs, value-added factors and measurable outputs from which a **correction**, a **corrective action**, or a **preventive measure** (see § 12.2, 12.3 and 12.4), and therefore an improvement, is possible (see figure 16.1).

It is therefore necessary to emphasize the identification of processes whose efficiency can be improved and to seek improvements by:
- Control of the process with a measurement of the results (quality indicators),
- Process analysis,
- Identification of the nonconformities, failures, non-quality-related costs, etc. in relation to the needs,
- Investigation of the cause(s) of nonconformities,
- Measurement of the personnel and customer satisfaction,
- Corrections, and planning of corrective and preventive actions.

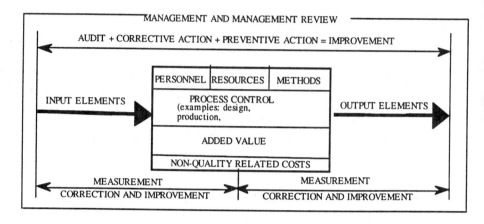

Figure 16.1 Approach for process improvement

In accordance with the ISO 9001 and 9002 standards, the confidence is built on the "monitoring" of the process throughout its development; it may be continuously corrected (central part of figure 16.1). In accordance with ISO 9003, the confidence is only built on the final measurement.

A number of methods and tools used to resolve problems can contribute to rendering this approach efficient

16.4 Methods, tools and techniques

Within the development of quality control concepts, and subsequently of quality management and total quality management (see § 1.4), a certain number of methods, tools and techniques contributing to the improvement of quality should be introduced progressively:
- Pareto diagram
- Ishikawa diagram (or cause-and-effect diagram),

Seven new quality management tools can be introduced to complete
the first generation of tools:
- affinity diagram
- relations chart,
- tree diagram,
- matrix diagram,
- arrow diagram
- decision chart (Process Decision Program Chart or PDPC)
- data analysis.

Figure 16.2 gives a list of tools that are described in the ISO 9004-4
draft standard.

The new tools should be implemented in the total quality management
methods such as:
- Quality function deployment - (QFD), applied to complex product
  development processes;
- Policy deployment.

16.5 Towards excellence

The continuous quality improvement leads towards excellence; but
excellence is not the object of standards! For an organization, excellence as far as quality is concerned is the total satisfaction of all the
stated and implied needs of all its members and customers. It is of a
subjective nature, but it can be the object of a certain evaluation carried out through the measurement of several management factors and
through several significant results, in relation to well-chosen excellence criteria.

In keeping with this spirit that "quality awards" have been instituted,
such as the Deming Award in Japan, The Malcolm Baldrige Award in
the United States, the European Award and the French Quality Award
(with the help of the Ministry of Industry). To qualify for the last two
awards, applicants must merit 300 of the potential 1000 points outlined by ISO 9000 standards that are required for these awards.

Tools and Techniques	Applications for quality improvement
1. Data collection form	Systematically gather data to obtain a clear picture of the facts.
Tool and technics for non-numerical data	
2. Affinity diagram	Organize into grouping a large number of ideas, opinions or concerns about a particular topic.
3. Benchmarking	Compare a process against those of recognized leaders to identify opportunities for quality improvement.
4. Brainstorming	Identify possible solutions to problems and potential opportunities for quality improvement.
5. Cause and effect diagram (Ishikawa)	• Analyze and communicate cause and effect relationships. • Facilitate problem solving from symptom to cause to solution.
6. Flowchart	• Describe an existing process. • Design a new process.
7. Tree diagram	Show the relationships between a topic and its component elements.

Tools and techniques for numerical data	
8. Control chart	• Diagnosis - evaluate process stability. • Control - determine when a process needs to be adjusted and when it needs to be left as it is. • Confirmation - confirm an improvement of a process.
9. Histogram	• Display the pattern of variation of data. • Visually communicate information about process behavior. • Make decisions about where to focus improvement efforts
10. Pareto diagram	• Display in order of importance the contribution of each item to the total effect. . Rank improvement opportunities.
11. Scatter diagram	• Discover and confirm the relationships between two associated sets of data. • Confirm the anticipated relationships between two associated sets of data.

Figure 16.2 Example of tools and techniques for quality improvement
(*Source:* ISO 9004-4)

Figure 16.3 shows a model adopted for the European Award. Note that the box "process" corresponds effectively to the share of the quality system elements that are the object of the ISO 9000 standards.

Finally, note that the example of quality manual structure of Figure 3.5 is presented (see § 3.3.2) by analogy with such a model that. such a structure - Management/Process/Results - contributes to ease the implementation of provisions that contribute to quality improvement. It can also facilitate the evaluation of an organization in its approach to obtain a quality award.

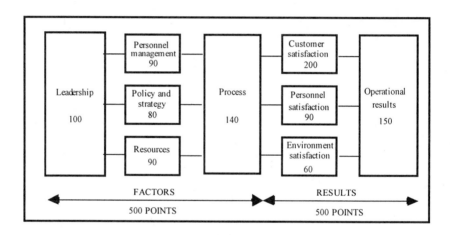

**Figure 16.3   Model of an organization adopted for
European Quality Award**
(*Source:* EFQM)

Conclusion

In the introduction to this book, attention was drawn to the role and importance of the quality manual within a quality management approach.

If the main theme of this book is considered again, from trhe definition of quality management through the establishment of a system for quality assurance concepts, it can clearly be understand why the quality manual is such a strategic tool for an approach to quality

An organization (enterprise, administration, etc.), preparing to draft a quality manual should:
- start by demystifying the quality management concepts so that they can be used as tools of efficiency and improvement: for management, which defines the quality policy of the organization as well as for those who, from the highest to the lowest level of hierarchy, will implement this policy; it is necessary that the language used in this policy be understood by everyone;
- think about the place of the organization in its commercial, industrial or administrative environment in relation to its activities, resources and level of technical progress by responding to the demands of all the parties: satisfaction of the customers, personnel, regulatory requirements, shareholders, etc.;
- analyze the existing organizational structure in order to improve the efficiency, by describing clearly and by redefining, where applicable:
  . the processes concerned,

. the assignments of the personnel,

. the resources and methods used;
- implement a documentary structure known by all and of which the quality manual will be the "head" document and a reference tool, in order to either:

    . establish the useful and efficient functional documents,

    . eliminate useless documents;
- establish reference tools in order to:

    . plan the quality in accordance with the demands,

    . control the processes (from the market and environment analysis to the design, production and the end of the life-cycle of the product or activity) and to reduce the incidence of corrective and improvement actions.

It is desirable that the drafting of the quality manual not be motivated by an obligation: from a client, from a standard of required application (for the certification of the quality system, for example). It is preferable that the drafting be carried out in a voluntary manner, in the spirit of quality management and of a progressive approach. Therefore, the ground should be prepared for a continuous quality improvement, leading towards excellence, in competitive international markets. The certification of the quality system, where applicable, or certain "Quality Awards" can thus be obtained more easily.

In this spirit, the drafting of the quality manual should be an occasion for the management of the organization:
- to evaluate its position by referring to standardized quality criteria or in relation to the competitive market;
- to clarify its technical and commercial strategy, to deal with its customers as well as suppliers and to establish solid basis for a good relationship with its partners;
- to search for malfunctions and non-quality-related costs.

Finally, it should be borne in mind that the quality manual is part of an open-ended process and that it gives a "written picture" of the level obtained by the organization in its quality management approach. This approach can itself be extended to certain quality aspects, for example, those of safety and the protection of the environment. From the perspective of quality assurance, it is the essential tool used to gain the confidence of the customer within a competitive commercial environment.

Normative Documents

Basic concepts

NF EN ISO 8402	Quality management and quality insurance - Vocabulary (1995)
(ISO 8402: 1994)	
ISO/IEC Guide 2	General terms and their definitions concerning standardization and related activities (1991)
AFNOR XG50	Quality and management - Vision 2000: a strategy for international standards' implementation in the quality arena during the 1990s (Report of the ISO/TC 176 Ad Hoc Task Force on the ISO series architecture, numbering, and implementation - December 1991)
X 50-125	Management de la qualité et assurance de la qualité - Vocabulaire - Termes complémentaires (1995 - ISO 8402: 1994 and additions)
VIM	International vocabulary of basic and general terms in metrology (1993).

Quality management / General and economical aspects

NF EN ISO 9000-1	Quality management and quality assurance standards -
(ISO 9000-1: 1994)	Part 1: Guidelines for selection and use (August 1994)

NF EN ISO 9004-1	Quality management and quality system elements -
(ISO 9004-1: 1994)	Part 1: Guidelines (August 1994)

NF EN 29004-2	Quality management and quality system elements -
(ISO 9004-2: 1994)	Part 2: Guidelines for services (August 1993)

ISO 9004-3	Quality management and quality system elements - Part 3: Guidelines for processed materials (1993)

ISO 9004-4	Quality management and quality system elements - Part 4: Guidelines for quality improvement (1993)

ISO 9000-4	Quality management and quality assurance standards -
IEC 300-1	Part 4: Guide to dependability management (1993)

X 50-126	Gestion de la qualité - Guide d'évaluation des coûts résultant de la non-qualité (October 1986)

X 50-127	Gestion de la qualité - Recommandations pour obtenir et assurer la qualité en conception (January 1988)

X 50-128   Gestion de la qualité et éléments de système
                                        qualité - Lignes directrices pour les achats et
                                        les approvisionnements (December 1990)

X 50-180-1   Qualité et Management - Défauts de contribu-
                                        tion du compte d'exploitation pour l'indus-
                                        trie et les services - Partie 1: Identification
                                        de la réserve cachée de productivité pour les
                                        achats et les approvisionnements (April
                                        1994)

Quality assurance

NF EN ISO 9001   Quality systems - Model for quality assurance
(ISO 9001: 1994)   in design, development, production, installa-
                                        tion and servicing (August 1994)

NF EN ISO 9002   Quality systems - Model for quality assurance
(ISO 9002: 1994)   in production, installation and servicing
                                        (August 1994)

NF EN ISO 9003   Quality systems - Model for quality assurance
(ISO 9003: 1994)   in final inspection and test (August 1994)

ISO 9000-2   Quality management and quality assurance
                                        standards - Part 2: Generic guidelines for the
                                        application of ISO 9001, ISO 9002 and ISO
                                        9003 (1993)

ISO 9000-3   Quality management and quality assurance
                                        standards - Part 3: Guidelines for the applica-
                                        tion of ISO 9001 to the development, supply
                                        and maintenance of software (December
                                        1991)

Quality audits

ISO 10011-1   Guidelines for auditing quality systems -
                       Part 1: Auditing (1990)

ISO 10011-2   Guidelines for auditing quality systems -
                       Part 2: Qualification criteria for quality sys-
                       tems auditors (1991)

ISO 10011-3   Guidelines for auditing quality systems -
                       Part 3: Management of audit programmes
                       (1991)

Tools / Methods / Techniques

X 06-011   Statistique et qualité - Définition, choix et
                       spécification de la qualité d'une fourniture de
                       produits industriels (April 1993)

X 06-012   Statistique et qualité - Méthodes de contrìle de
                       la qualité technique d'une fourniture de pro-
                       duits industriels (April 1993)

NF X 07-10   Métrologie - La fonction métrologique dans
                       l'entreprise
                       (December 1992)

X 07-11   Métrologie - Constat de vérification des moy-
                       ens de mesure
                       (October 1992)

X 07-15   Métrologie - Essais - Métrologie dans
                       l'entreprise - Raccordement des résultats de
                       mesure aux étalons (December 1993)

X 07-16	Métrologie - Essais - Métrologie dans l'entreprise - Modalités pratiques pour l'établissement des procédures d'étalonnage et de vérification des instruments de mesure (December 1993)
NF X 50-142	Relations clients-fournisseurs - Qualité des essais - Lignes directrices pour demander et organiser les essais (December 1990)
X 07-143	Métrologie - Essais - Conception et réalisation des essais - Pertinence et représentativité des essais (December 1990)
ISO 10005 (Ex ISO 9004-5)	Quality management and quality system elements - Part 5: Guidelines for quality plans (1995)
ISO 10007 (Ex ISO 9004-7)	Quality management and quality system elements - Part 7: Guidelines for configuration management (1995)
ISO 10013	Guidelines for developing quality manuals (1995)
NF X 50-160	Gestion de la qualité - Guide pour l'établissement d'un manuel qualité (October 1988)
X 50-161	Manuel qualité - Guide pour la rédaction d'un manuel qualité (December 1988)
X 50-162	Relations clients-fournisseurs - Guide pour l'établissement d'un manuel assurance qualité (October 1991)

X 50-163	Qualité et management - Typologie et utilisation de la documentation décrivant les systèmes qualité (December 1992)
NF X 50-164	Relations clients-fournisseurs - Guide pour l'établissement d'un plan d'assurance qualité (June 1990)
X 50-165	Page de garde de rapport d'audit qualité (January 1989)
X 50-168	Relations clients-fournisseurs - Questionnaire type d'évaluation préliminaire d'un fournisseur (October 1988)
X 50-170	Qualité et Management - Diagnostic Qualité (December 1992)
X 50-171	Qualité et Management - Indicateurs et tableaux de bord qualité (October 1993)
NF EN 30012-1	Quality assurance requirements for measuring equipment -
(ISO 10012-1: 1992)	Part 1: Metrological confirmation system for measuring equipment (February 1994)
RG Aéro 00040	Recommandation générale pour la spécification de management de programme, BNAE (December 1991)

Quality and Management - ISO drafts

| ISO/CD 10006 | Quality management - |
| (Ex ISO/WD 9004-6) | Part 6: Guidelines to quality in project management (1995) |

ISO/CD 10014   Guidelines for managing the economics of
                      quality

ISO/CD 10012-2   Quality assurance requirements for measuring
                      equipment -
                      Part 2: Measurement assurance

Value Analysis - Functional analysis

NF X 50-150   Analyse de la valeur - Analyse fonctionnelle -
                     Vocabulaire
                     (August 1990)

NF X 50-151   Analyse de la valeur - Analyse fonctionnelle -
                     Expression fonctionnelle du besoin et cahier
                     des charges fonctionnel (December 1991)

NF X 50-152   Analyse de la valeur - Caractéristiques fon-
                     damentales
                     (December 1991)

NF X 50-153   Analyse de la valeur - Recommandations pour
                     sa mise en oeuvre
                     (May 1985)

Maintenance

X 60 151   Maintenance industrielle - Entreprises presta-
                     taires de service - Guide d'application des
                     normes ISO 9001 - 9002 et 9003
                     (August 1993)

X 60 200   Documents techniques à remettre aux utili-
                     sateurs de biens durables à usages industriels
                     et professionnels - Nomenclature et principes

généraux de rédaction et de présentation
(April 1985)

X 60 500   Terminologie relative à la fiabilité - Mainten-
                        abilité - Disponibilité (October 1988)

X 60 502   Fiabilité en exploitation et après-vente
                        (December 1986)

X 60 510   Technique d'analyse des systèmes (December
                        1986)

X 60 520   Prévision des caractéristiques de fiabilité
                        (May 1988)

Certification

NF EN 45001   General criteria for the operation of testing
                        laboratories
(EN 45001: 1989)   (December 1989)

NF EN 45002   General criteria for the assessment of testing
                        laboratories
(EN 45002: 1989)   (December 1989)

NF EN 45003   General criteria for laboratory accreditation
                        bodies
(EN 45003: 1989)   (December 1989)

EN 45004   General criteria for the operation of bodies per-
                        forming inspection (1995)

NF EN 45011   General criteria for certification bodies operat-
                        ing product
(EN 45011: 1989)   certification (December 1989)

NF EN 45012   General criteria for certification bodies operat-
                             ing quality system
(EN 45012: 1989)   certification (December 1989)

NF EN 45013   General criteria for certification bodies operat-
                             ing certification of
(EN 45013: 1989)   personnel (December 1989)

NF EN 45014   General criteria for suppliers' declaration of
                             conformity
(EN 45014: 1989)   (December 1989)

Environmental management

X 30-200   Système de management environnemental
                             (Experimental standard - April 1993)

X 30-201   Lignes directrices pour l'audit des systèmes de
                             management environnemental (January 1994)

X 30-202   Critères de qualification pour les auditeurs de
                             systèmes de management environnemental
                             (January 1994)

X 30-203   Management des programmes d'audit de
                             systèmes de management environnemental
                             (January 1994)

ISO/CD 14000   Environmental management systems - General
                             guidelines on principles,
                             systems and supporting techniques (Draft 1994-
                             09-28)

ISO/CD 14001   Environmental management systems - Specifi-
                             cation with guidance for use (Draft 1994-09-28

	- With correspondence with ISO 14001 and ISO 9001)
ISO/CD 14010	Guidelines for environmental auditing - General principles of environmental auditing (Draft 1994-05-30)
ISO/CD 14011-1	Guidelines for environmental auditing - Audit procedures - Part 1: Auditing of environmental management systems (Draft 1994-05-30)
ISO/CD 14012	Guidelines for environmental auditing - Qualification criteria for environmental auditors (Draft 1994-05-30)
ISO/CD 14021	Environmental labeling - Self-declaration environmental claims - Terms and definitions
ISO/CD 14040	Life cycle assessment - General principles and practices

Index